THE WESTERN WAY

To my wife, Nikki, for her endless patience, wit, and encouragement.
To Ewan Lawson, for inspiration, needling, challenge, and humour.
And to the RUSI Military Sciences research group of the day: Pete Quentin,
Elizabeth Braw, Sid Kaushal, Sarah Ashbridge, Jack Watling, Peppi Vanaanan,
Justin Bronk, Ali Stickings, Nick Reynolds, Julianna Suess, and Andy Young.
A grand bunch of people to work with: endlessly stimulating.

The Western Way of War

PETER ROBERTS

Howgate Publishing Limited

First published in 2024 by
Howgate Publishing Limited
Station House
50 North Street
Havant
Hampshire
PO9 1QU
Email: info@howgatepublishing.com
Web: www.howgatepublishing.com

British Library Cataloguing-in-Publication Data
A catalogue record for this book is available from the British Library

ISBN 978-1-912440-50 4 (pbk)
ISBN 978-1-912440-51 1 (hbk)
ISBN 978-1-912440-62 7 (ebk – ePUB)

The views expressed in this publication are those of the individual authors and do not necessarily reflect official policy or position.

Contents

Foreword

We in the West have become devoted to our 'Way of War' simply because it has worked in the past. It succeeded in building empires and winning world wars and has dominated military prowess for centuries; and being wedded to it will, if I am any judge, prove to be our undoing.

That is why this book matters. Look at the world today; it is uncertain and unpredictable, and, despite our best efforts, it is moving away from us. New powers and old empires are today set against the western way of life. These nations are aligned towards the common goal of disrupting and diminishing our authority and influence on the world stage.

Social media shapes everyone's opinion; disinformation provides an alternative reality; unworthy characters amongst many of our leaders allied with their selfish self-interest, corrupts confidence in our national and international institutions. Dangerous alternative and credible choices are increasingly taking shape and threaten our way of life while we remain blindly confident that our 'Way of War' will safeguard our interests. This is a false God. Only by understanding our 'Way of War' can we start to understand the clear and present danger we currently face. These alternative regimes, through their global cooperation's, technical and financial institutions, political and diplomat engagements, military and para-military forces, are all underpinned by active propaganda and disinformation which represent a very different 'Way of War'.

The 'Western Way of War' will likely prove to be our downfall and that is why this is more than just an interesting and informative read. It is as Sir Winston Churchill so timely put it – *'this is not the end, it is not even the beginning of the end but it is perhaps the end of the beginning'* of the 'Western Way of War'.

General Sir Graeme Lamb KBE, CMG, DSO

Introduction

One can write entire books trying to define what a *way of war* is, how it is tied to the strategic culture of a state, or a nation, and how reflective it is of the military thinking of any particular state:[1] indeed some have at length, or in more pithy formats.[2] Yet for the purpose of practitioners in the profession of arms (the audience for this book), a way of war (and warfare), is the chosen means by which a state fights – in essence, how a state or nation decides *how* it will engage in combat. It will not necessarily determine *what* it will fight over, although some states have provided an indication of that; and when they do, it necessarily becomes part of the idea of their way of war. Rarely do such statements define what values or interests that it will most definitely fight over, but rather the *how* refers to the means it will employ to do so when such matters are determined worthy, and how they will be employed. Historically, the way a nation engages in conflict was based on a community of knowledge given their unique history, culture, moral values and ethics, force available at hand (and from allies), a determination of their values and interests, and – critically – the adversary they were facing. Key within this rubric has always been the decision-making element of going to war, both politically and militarily. Given how states had such different experiences in war and warfare it will be no surprise that considerable differences have existed in the way in which they fought. Yet the thesis of the term 'Western way of war' is that states of the West have surrendered their national differences and adopted a broadly single idea of how they will engage in contests of arms.

In using the term *The West*, it is acknowledged that this taxonomy is a sloppy, inexact and sometimes pejorative grouping word for a collection of states. Depending on where you reside on the globe, west as a direction

---

[1] Azar Gat, *A History of Military Thought: From the Enlightenment to the Cold War* (Oxford University Press, 2011).

[2] Jack L. Snyder, *Soviet Strategic Culture: Implications for Limited Nuclear Operations* (RAND Corporation, 1977); Gregory Vincent Raymond, *Thai Military Power: A Culture of Strategic Accommodation* (NiAS Press, 2018).

can mean a variety of things. Some scholars have given the term a good deal of thought and provided complex definitions after a good deal of hand wringing.[3] In Europe, *the* West means European, plus the North Americans. Antipodeans would be included in the group after cross examination – a deeply unfair afterthought given the role that Australians are playing in thinking about conflict and war today. After yet more nudging, Europeans might also include Japan, South Korea and perhaps even India too. It fast becomes a term that encompasses all liberal, or semi liberal democracies. But really it is about 'people like us': Allies of the United States of America, who seem to hold broadly similar values. As a wider description, it also includes some Caribbean, Latin American, African and Asian states too. Perhaps, at its widest possible boundaries, it has a negative meaning: if you are not *The West* you are an adversary, or at least a potential enemy. Given that descriptor, the club of the West is staggeringly large. For such a wide church, one would expect their militaries and the doctrine on which they fight to be equally as diverse and disparate.

For those with experience of war and warfare, it would perhaps seem logical that militaries would fight in a different way for each war they became engaged in (whether a deliberate decision or not). Since the timing, geography, economic means, and adversary of each war is remarkably different from any other, one might imagine that using the same procedures, tactics and processes for each campaign would be context specific and would need to alter – perhaps radically – between each combat operation. One would not expect to use the same doctrine in fighting terrorism or an insurgency, as one would in fighting an armed invasion by conventional military forces. The approach of a military force against an armed militia in Libya would be ineffective against a Chinese armoured division attacking India.

Yet since the 1980s, democratic states have become rather fixated on determining and divulging which values and interests they deem worthy of fighting over,[4] and in dictating a single method of combating every challenge. The policies were widely shared in speeches by political leaders,

---

[3] William McNeil, 'What We Mean by the West', Western Civilisation in World Politics, *Orbis*, Fall 1997, 513-524 (https://www.fpri.org/wp-content/uploads/2016/07/WH-McNeil-What-We-Mean-by-the-West.pdf).

[4] See for example, Frank Hoffman, 'A Second Look at the Powell Doctrine', *War On The Rocks*, 20 February 2014, https://warontherocks.com/2014/02/a-second-look-at-the-powell-doctrine/; or the ill-fated 'Blair Doctrine', defined in a speech made by then British Prime Minister Tony Blair in Chicago on 22 April 1999, https://archive.globalpolicy.org/component/content/article/154-general/26026.html.

and their methods of fighting published on the internet.[5] None of this was a secret it seemed.

Such proclamations might have been helpful in providing clearer direction to militaries involved in planning what their forces needed to look like, but it had the distinct disadvantage in signposting to adversaries how Western militaries would engage in every contest. Both these factors caused problems when Western militaries did actually engage in combat operations: because their structures, designs and ways of fighting did not align to how adversaries behaved, but also in that they had become predictable and bureaucratic.

This is the reason that discussions over *how* a military fights – its way of war and warfare – is so important and relevant today. Important, in that fighting causes death and destruction, and costs blood and treasure; relevant because it seems that discussions over ways of war and warfare have been much neglected by those engaged in the Profession of Arms.

It was not a surprise – disappointingly – that in starting a discussion over the Western way of war in a series of podcasts in 2020 few senior serving military personnel were willing to engage in a discussion in a meaningful way (for example, one that departed from a pre-authorised script and was open to challenge). The topic was, to many, also considered irrelevant to the matters of the day.[6] Two years later, a renewed interest had been triggered: because of how an adversary was fighting, not because of the short-sightedness of Western policy makers and military leaders.[7]

A Western Way of War?

How a military fight, indeed how a state fights is perhaps the greatest indicator of whether it will achieve success on the battlefield. That idea of

---

[5] See, for example, the release of NATO EXTAC 1000 series 'Maritime Maneuvering and Tactical procedures", released in 1996 http://nato.radioscanner.ru/files/article66/1000.pdf, or NATO ATP-3.2.1 'Allied Land Tactics', https://moodle.unob.cz/pluginfile.php/98398/mod_resource/content/1/ATP-3.2.1%5B1%5D%20ALLIED%20LAND%20TACTICS%20 2009.pdf.

[6] Discussions and correspondence with the author from more than 20 senior military officers from across Western militaries between June 2020 and December 2021.

[7] This refers to the Russian invasion of Ukraine on 24 February 2022, with conventional forces. This was a very different method to what had been predicted and expected about the conduct of war by Western militaries and the mainstream popular media. See, for example, Amos Fox, 'Reflections on Russia's 2022 Invasion of Ukraine', Land Warfare Paper 147, Association of the United States Army (AUSA), September 2022.

how is not simply tactics, but also the platforms, the people, the culture, the societal support, the will to fight, the economic means to pursue a campaign, the support from allies, organisation and institutions and a myriad of other factors. But determining what this *how* looks like, or measuring it for different states, is perhaps unmanageable. One might search military or national doctrine for a definition, but it rarely appears in a codified structure or policy document. In researching the idea of 'how' states fight (and it differs considerably between societies as it always has), one needs to examine the history of a state, but also the narratives of governments and controlling entities, the speeches of key leaders and policy makers, as well as the investment decisions that nations make.

Yet such scholarly research will only provide a highly retrospective view of how a state intends to engage in armed contests. Sometimes, leaders will let slip what they think in more private discussions; some scholars have even put this in print.

Much of the discussion around a Western way of war has been, across literature and academic study, rooted in the Greek experiences as codified by Aristotle and Heroditus. Their distinct, and somewhat amusing, opinion on the naturally inbred superiority of Greeks as great fighters was based largely on the experiences of conflict against the Persians, rather than internal civil wars between the Greek city states. Codified by Victor Davis Hanson, the idea that Greeks always sought direct confrontation, a single big battle with a decisive outcome and over which matters of importance could be settled, was contrasted with a view that everyone but the Greeks fought in an indirect manner.[8] This deeply jingoistic view of a nationalistic way of war has been widely replicated as a state approach to combat and warfare: the French adopting much the same language and cultural exceptionalism as was present in the writing of Aristotle and Heroditus.[9]

Much of the discussion around the Greeks can be rubbished. There is evidence enough to debunk the myth that Greeks always sought direct confrontation in Homer's epic, *The Iliad*, where the wit, guile, cunning and ruse of Odysseus were the instruments of Trojan defeat, not direct engagement.[10] Yet there are facets of Greek fighting that do seem to remain

---

[8] Victor Davis Hanson, *The Western Way of War: Infantry Fighting in Ancient Greece* (University of California Press, 2000).
[9] See, for example, the work of Christine de Pizan in articulating the elan and martial virtues of France in comparison to neighbouring states. Christine de Pizan and Angus Kennedy, *The Book of the Body Politic* (Iter Press, 2021).
[10] Bernard Knox Homer (ed), *The Iliad* (Penguin Classics, 1992).

nested into a Western style of war. Just as Greeks imposed a values-based system that developed the just war tradition (popular in Western mantras, to varying degrees over time, about justification for conflict), so too do the ideas of valour and personal courage sit at the heart of Western military philosophies. And while it might be disputed, there remains a deep-seated belief in Western military planning assumptions about bringing about a culminating battle over which vital matters might be decided upon.

In dealing with *how* the West fights today, it is of course acknowledged that by the 19th Century there were several historical schools of military theory: Prussian, French, British, Russian, Italian and Japanese – to name but a few – which were all differentiated and individually recognisable.[11] These have been identified as peculiar to those states, imbued with some of the core cultural phenomena of those indigenous people, and the deliberate changes made to their military practices and institutions on the basis of their own discrete experiences in conflict, campaigns, personalities, and warfare as lived – rather than as a collective that was able to learn the fullest gamit of lessons from each other. Arguably, these merged into a single school by 1990: An American led doctrine and concept of fighting that emerged from the Cold War and that was centred on the core belief that technological superiority could overcome the mass of the Warsaw Pact forces. It is perhaps a little tragic that so many of the previous lessons and individual schools of military theory have all but disappeared.

Since 1990, and the move towards a single Western way of war heralded by the arrival of Air Land Battle doctrine[12] from the United States (and rapidly adopted by NATO as the common Alliance doctrine for fighting), there has been a peculiar obsession with technology: a deep-seated belief that the belligerent with the most technologically advanced military capabilities was pre-ordained to be victorious. There is little evidence for this assumption,[13] yet it remains the pre-eminent assumption and foundational principle of contemporary military theory across the West. The UK is an interesting place to observe this from.

---

[11] JJ Widen, *Contemporary Military Theory* (Routledge, 2013).

[12] Douglas Skinner, 'Air Land Battle Doctrine', Center for Naval Analyses, Professional Paper 463, September 1988, https://apps.dtic.mil/sti/pdfs/ADA202888.pdf.

[13] See, for example, John France, *Perilous Glory* (Yale University Press, 2013), which provides a glorious overview of the evolution of warfare from ancient times to modern warfare, covering nearly every civilisation since the beginning of recorded history.

Time, Space and Context

At an off-the-record, closed-door roundtable at RUSI in 2019, a former British Vice Chief of Defence Staff stated that, "Information will be the decisive domain in the next war". This conclusion came on the back of costly in-house research from a dedicated centre (the UK Ministry of Defence's Defence, Concepts and Doctrine Centre)[14] that was tasked with laying out what the future of war would look like. In a pre-2022 world, apparently the dominant – perhaps only – future battlefields would be in man-made cyberspace and would not involve actual death and destruction. This seemed somewhat at odds to the lived experiences of those actually engaged in contemporary conflict, certainly it was a different storyline to the ones used after Russia's second invasion of Ukraine in February 2022 using conventional forces. The descriptions used then appeared to overturn some of the core truisms of war and warfare: most importantly, conflict cannot exist without human imperil.[15]

In this, the British military memory seems somewhat short lived.[16] Between 2017 and 2022, the British military and political establishment openly and completely embraced the commercial ideas of war and warfare being determined by information and data, with an associated narrative that elevated information, and the impact of cognitively influencing an adversary, above all other elements – including the platforms, equipment, people and weapons actually required to wage war.

The UK was not alone in this regard. Increasing numbers of military leaders from liberal democracies seem to have been persuaded that messaging alone could manipulate and coerce an enemy (leaders, soldiers or hostile publics), and *ergo* that war can be won through warfare waged without any cost of blood or treasure. As European states continued to divest themselves of fighting platforms, weapons and people, one might wonder where the evidence of these presumptions lay? It is a peculiar facet to the argument that even leading technologists did not go as far as political and military leaders of the time. [17]

---

[14] *Global Strategic Trends (2018)*, Defence Concepts and Doctrine Centre, UK Ministry of Defence, 2018. https://assets.publishing.service.gov.uk/government/uploads/system/uploads/attachment_data/file/1075981/GST_the_future_starts_today.pdf.
[15] Paddy Walker and Peter Roberts, *Wars Changed Landscape?* (Howgate, 2023)
[16] Jim Storr, *Hall of Mirrors* (Helion and Company, 2019).
[17] Shyam Sankar, 'Data is the new snake oil', St Gallen Business Review, 9 May 2019, https://www.stgallenbusinessreview.com/data-is-the-new-snake-oil/.

In fact, the evidence on the influence of cognitive coercion in warfare pointed directly in other direction. Failed strategies of coercion and decapitation (or regime removal in modern parlance), litter contemporary history of the day. The romanticised notion of bloodless victories through cyber, information campaigns, and an almost mystical use of data (sometimes incorrectly referred to as Artificial Intelligence), was – and continues to be – fiction. The evidence from large wars Iraq, Afghanistan, Syria, Ukraine, and Yemen bear witness to this.

Just as concerning was and remains the demand from commanders for 'speed' in everything, particularly in decision-making. Again, the assumption was – and continues to be – that more data will enable this. Yet speed in itself is not necessarily helpful: it does not mean making the correct decision, indeed, not even better decisions. The research in this also differs from expectations.[18] Speed in and of itself is not a helpful metric in successful military operations, nor in strategy. If one wishes to make the right decision, then more uninterrupted thinking space is needed. The power and science of boredom generates more neural connections to be made and greater clarity to be delivered: in turn leading to better decisions (this is why people exercising, playing sport, reading books, or engaged in box-set marathons often come to a 'Eureka' moment during the periods in which they are free from the distractions of modern life).[19] Subconsciously perhaps, militaries around the world reject this with their own experiences instead determining that constant connection to a command and control 'singularity' will somehow – miraculously – bring about kingfisher moments of fleetingly brilliant insight. It took Marcus Aurelius many years of austere campaigning to come to conclude that a different approach was needed,[20] but perhaps todays wiki-commanders now leading liberal militaries have given this a good deal more deep reflection and are thus, somehow, wiser than their predecessors across human history?

Instead, the evidence indicates that less data, more decisions – but at a slower speed, aligned with greater delegation – are important.[21] But none of these are as vital as the political will to use force, and the means to do so.

---

[18] Daniel Kahnerman, *Thinking, Fast and Slow* (Penguin, 2016).
[19] Mark Hawkins, *The Power of Boredom* (Cold Noodle Creative, 2016).
[20] Martin Hammond, *Marcus Aurelius: Meditations* (Penguin Classics, 2006).
[21] Kaknerman (2016), *op cit.*

The British Way of Warfare

The British (indeed Western) way of warfare is not codified in any meaningful way.[22] If speeches by politicians and military leaders are to be believed, then today's unwritten British way of warfare should be characterised as information centric, aligned with speed of decision-making, and a belief that victory can be achieved against an adversary with minimal casualties if fought at range (from the UK homeland), with technological superiority, and underpinned by brilliant command.[23] These factors, it seems, if brought together properly provide a pre-ordained right to victory.

The evidence makes a mockery of such assumptions, both historically and in terms of contemporary conflict. The assumptions and principles outlined above take no heed of the actions, activities, desires or fighting style of adversaries.[24] They emphasise speed of decisions not speed of military action – to shock and paralyse adversaries (a foundational facet of military success). They ignore the fact that belligerents with better tech rarely are victorious, promises to remove the fog of war are usually ill-found, and the myth of speed (versus good) decision-making litter history, recent and more distant.[25]

---

[22] See for example, the vacuous British Defence Doctrine hastily reissued in 2022, https://assets.publishing.service.gov.uk/government/uploads/system/uploads/attachment_data/file/1118720/UK_Defence_Doctrine_Ed6.pdf, in contrast to its predecessor https://assets.publishing.service.gov.uk/government/uploads/system/uploads/attachment_data/file/389755/20141208-JDP_0_01_Ed_5_UK_Defence_Doctrine.pdf.

[23] See for example, a UK Chief of Defence Staff annual lecture at RUSI. "The pervasiveness of information and the pace of technological change are transforming the character of warfare and providing new ways to execute this form of authoritarian political warfare including information operations, espionage, assassinations, cyber, the theft of intellectual property, economic inducement, the utilisation of proxies and deniable para military forces, old fashioned military coercion, using much improved conventional capability, and, of course, lawfare – all of which is backed by clever propaganda and fake news to help justify these actions." https://www.gov.uk/government/speeches/chief-of-the-defence-staff-general-sir-nick-carters-annual-rusi-speech.

[24] It is worth contrasting here the exchange on 17 November 2021 between UK PM Boris Johnson and Chair of the House of Commons Defence Select Committee at the Liaison Committee on national security challenges and capabilities needed to defend against them as an indicator of the divergence in understanding of war by politicians at the most senior level in the UK, Q139-Q149 at https://committees.parliament.uk/oralevidence/3007/default/, with a speech by former US Defence Secretary James Mattis at the OSS as reported by Walter Pincus https://www.thecipherbrief.com/column_article/general-james-mattis-and-the-changing-nature-of-war, in which General Mattis articulates a far more nuanced and contextual understanding of threats of the day.

[25] Mike Martin, *How to Fight a War* (C Hurst and Co, 2023).

The presentism and neophilia of today's military and political leaders – obsessed with the faddism of innovation, transformation, data, digitisation, AI, autonomy, 'speed of relevance', and robotics free from human control – needs a bold challenge to force the application of the corrective lens before disaster and defeat strike. It requires a new doctrine for the British way of warfare.

The British Way of Warfare: A History

In 1933 at a lecture at RUSI, Basil Liddel-Hart coined the term the British Way of War.[26] His thesis, founded on his experiences of World War I, was that some principles had emerged in the way that Britain – and her dominions of the time – fought wars, wars of all types including colonial, coercive, insurgency, grand or limited in scope. Those core ideas were shaped around an 'indirect approach', and 'limited liability'. In essence, Liddel-Hart believed that Britain was constantly choosing between two diametrically opposed aims: either in providing and influencing a balance of power in continental Europe with land forces (armies), or alternatively ensuring the freedom of access, trade routes and regional security and stability in maritime power (naval power). In this argument, Liddel-Hart established the British way of war as a choice between the prioritisation of the armed forces between land and sea, which should be the preeminent concern, and politically he charged the policy makers with deciding what mattered more and funding it – Europe or the Rest of the World. It was this sort of choice that Liddel-Hart underpinned as the British way of warfare.

In 1972 Michael Howard critiqued Basil Liddel-Hart's thesis in a reappraisal of the idea of a British way of war over a far longer span of history.[27] Howard, with his usual prescience, pointed out that Liddel-Hart was trying to solve 20th Century problems through the application of 18th Century 'recipes'. He noted that in advocating for a maritime strategy as the British way of war the result would be to effectively allow continental allies to be defeated since British land forces would not be present at their first battle and therefore this engagement could, perhaps, become the decisive

---

[26] Later reproduced as an essay in the RUSI Journal. Basil Liddel-Hart, 'Economic Pressure or Continental Victories', *RUSI Journal*, vol. 76, no. 503 (1931), 489-510.
[27] Michael Howard, *Continental Commitments: Dilemma of British Defense Policy in the Era of the Two World Wars* (Prometheus Books, 1972).

battle (since without the added strength of the British and Imperial forces, continental armies would be over matched in terms of quantity and quality of an adversary's forces). Howard thought that in adopting a maritime strategy, Britain was selecting an option that – on the basis historical precedent – was less than optimal, presenting, as it did, an almost unbroken record of expansive and humiliating failures of British maritime power, covering almost a century of time.[28] The side effect of such a policy was in a loss of 'leverage' and influence in the politics and events on the European continent. Howard's re-examination of the challenges concluded that the preference for a peripheral strategy – one that played around the edges, the seams, rims of the problem through the application of maritime power – side lined the greatest risk to national security. The prominent historian noted that the choice between a maritime strategy and a continental one, as presented by Liddel-Hart, was not a set of alternatives but instead these should be regarded as complementary and interdependent aspects of the same political desire.[29]

Hew Strachan conducted a further assessment of the British way of war in 1981, deep in the midst of the Cold War. With a commitment to the protection of continental Europe alongside NATO in the contest against the Warsaw Pact, the UK had prioritised defence (and foreign) policy towards land forces prepositioned in Europe. Backed by a gradual but continual shrinking of its economic position globally, and by the reduction in overseas commitment as the British Imperial stakes abroad declined, the requirement for a significant naval force waned to one that prioritised protection of the European northern flank and shipping lanes from the United States. Thus by this stage, observed Strachan, that much of the debate on Britain's strategic position, "rested on the assumption of military inevitability. British defence policy over the previous 100 years was presented as a steady and intractably, if somewhat reluctantly, progressing towards its 'natural state', a continental commitment."[30]

Herein lies one of three valid critiques of an identification of the British way of war: that the recorded evidence of it is entirely shaped by the

---

[28] Michael Howard, 'The British Way in Warfare: A Reappraisal,' in Michael Howard (ed.) *The Causes of Wars and Other Essays* (Harvard University Press, 1983), 179.
[29] M Howard, *op cit* (1972), and again in Michael Howard, 'The British Way in Warfare: A Reappraisal,' in Michael Howard (ed.) *The Causes of Wars and Other Essays* (Harvard University Press, 1983), 170-173.
[30] Hew Strachan, 'The British Way in Warfare Revisited,' *The Historical Journal*, vol. 26, no. 2 (1983), 447-461.

politics and strategy of the day. The Cold War defined the required way of war of the day – as much for the US, Germany, and Norway, as it did for the UK.[31] This was equally true of an assessment of the British Way of War in the Hundred Years War, as it was for the Greeks during the Peloponnesian Wars. Indeed, it maybe this that caused Paul Cornish to comment that, "UK culture *shapes* but does not *direct* UK strategy".[32]

Britain's pre-eminent military historian, Lawrence Freedman, reappraised Liddel-Hart's thesis in a post-Cold War world in 2009 by examining the issue through the prism of Alliances.[33] He concluded that British strategy in a new era, in terms of the way of warfare, had come to depend on two key propositions. First that no serious international objective could be met without the aid of the United States. Second, that notwithstanding the first proposition, that the United States could not wholly be trusted to fulfil wider global policy objectives in a sensible and effective manner. This meant that the British were, in their own political interpretation of strategy in the day, required to influence America and American strategy with the benefit of British wisdom and experience! This position manifested itself in unclear core priorities – reflected in priorities, equipment programmes and budgets – but with a 'special relationship' between the UK and the USA sitting at the heart of any British national security discussion. Indeed, it seems that this facet remains true today as it did in 2009, with British national security budgets being prioritised into areas of capability that could add, in British minds, to the ability of the UK to leverage some kind of influence with the US political machine. It is perhaps the preeminent concept in the current incarnation of the British 'Way of Warfare', if indeed such a thing still exists today.

Common Conclusions and Principles

From a scholarly perspective, Liddel-Hart's key principles of the British way of war seem to have survived largely intact: economy of effort, the

---

[31] Colin McInnes, *Hot War, Cold War: The British Army's Way in Warfare 1945-1990* (Brassey's, 1996).
[32] Cornish, 'Strategic Culture,' 361 (original italics); K. Waltz, *Foreign Policy and Democratic Politics: The American and British Experience* (Longman, 1968), 7-8; also, Macmillan, 'Strategic Culture and British Grand Strategy 1945-1952', 36-37.
[33] Lawrence Freedman, 'Alliance and the British Way in Warfare'. *Review of International Studies*, vol. 21, no. 2, 1995, 145–58. JSTOR, http://www.jstor.org/stable/20097403.

indirect approach, and limited liability (in political terms). David French, writing on the same topic, highlighted three paradigms that encapsulated British strategic behaviour during the 12 wars it engaged in over the period 1945 to 1967: the peacetime paradigm, the wartime paradigm, and the mixed paradigm. Across these timelines, French highlighted economy of effort from this analysis, turning the unforgettable phrase that the British way of war was, "A calculus designed to achieve the dominant policy aims at minimum cost."[34] French added to Liddel-Hart's conclusions, from a military perspective, an approach focused on mobility and surprise.

It is, perhaps, the role of scholars to analyse and disagree. As such there is no single set of agreed principles on which a British way of war relies. It is, to coin a very academic term, an essentially contested topic:[35] no one agrees what it is, but everyone understands that it exists. As McInnes nicely articulates it, "The British way of war is a community of knowledge rather than abstract theorising."[36]

But it has to be acknowledged that the reliance on historical and Cold War tactics and processes by the British armed forces required that much of the developments in *how* they planned to fight was simply an adaptation of US principles – backward engineering of experiences in war and warfare to allow and optimise for military interoperability on an increasingly technical battlefield. This has not simply been the case for the British but has also been reflected in other NATO Allies – notably in Germany, the Netherlands, Denmark, Norway, Italy, and Austria, as well as Poland, the Czech Republic, and the Baltic states after the breakdown of the Warsaw Pact. Indeed, the increasing shift towards a single ubiquitous way of war across Western militaries has been notable since 2001, with even the French forces aligning themselves to US doctrine, processes and tactics.

Before leaving the theory and historiography of the British way of war as a term, it is worth highlighting the remaining two critiques of the term. If the first was that UK strategy, and thus its way of war, has always been defined by the politics and context of the day, then the second is that historians – as it is them who have been shaping the thinking about this

---

[34] David French, The British Way in Warfare, 1688-2000 (Unwin Hyman Inc., 1990), xii-xvii.
[35] W.B. Gallie, 'Essentially Contested Concepts,' *Proceedings of the Aristotelian Society*, vol. 56, (1955/56), 62-87, and William E. Connolly, *The Terms of Political Discourse* (Princeton University Press, 1983), 20. A discussion around the Aristotelian principle of essentialism, Gallie indicates there are 'concepts which are essentially contested, concepts the proper use of which inevitably involves endless disputes about their proper uses on the part of their user'.
[36] McInnes, The British Army, 127.

topic – have been engaged in advocacy as much as they have analysis.[37] The clear division between the 'blue-water' school of naval historians and those of the continentalists is evident from the passion with which they make their arguments, and their inability to seek or make compromise between their varied interests. The third critique is the relative immaturity of British strategy as a formal policy position in the UK. Britain only really started formal defence and security reviews in 1956, and articulating military doctrine in the 1990s, well before even the establishment of a National Security Council whose role it is to formulate such positions.[38]

War or Warfare?

Liddel Hart, and those who critiqued his paper, used the terms 'war' and 'warfare' as interchangeable. Famously, Carl von Clausewitz differentiated them: war as the grand strategic choices of policy, and warfare the practice of armed coercion and violence used to implement political strategy. Whilst academically pure, the reality is an overlap between these two spheres. While scholars pose important, grand strategic questions and want (perhaps even need) to separate them from reality, those engaged in the profession of arms need to understand the Western approach to warfare (how *we* fight, and how *adversaries* respond) as a critical military question.

Where possible in our podcast interviews, we adopted a definition of warfare as the practice of war – the practice of those verbs in Clausewitz's infamous definition – in a particular time, space, and context of fighting. It is the practice of violence and passion against other peoples that is the core nature of warfare. Thus, we asked guests to look at military ways of fighting, not the political construction of grand strategy.

---

[37] Andrew Lambert, 'The Naval War Course, Some Principles of Maritime Strategy and the Origins of "The British Way in Warfare," in Keith Neilson & Greg Kennedy (eds.), *The British Way in Warfare: Power and the International System, 1856-1956: Essays in Honour of David French* (Ashgate, 2010), 251.
[38] Brian Holden Reid, 'Introduction: Is There a British Military 'Philosophy'?,' in J.J.G. Mackenzie & Brian Holden Reid (eds.), *Central Region Versus Out of Area: Future Commitments* (Tri-Service, 1990), 1.

Boundaries for the Interviews

The interviews posed the question of whether there was, today, just a single school of way of warfare in the West. As a broadly accepted concept, it equally followed that the US school of warfare has been applied against all aggressors in roughly similar manners: counterterrorism, counter insurgency, high intensity conflict, civil wars, conventional deterrence, partnering and unlimited warfare. The core question was one of whether this single *Western way of warfare* was fit for task, and in any given context, because it has been applied as a single doctrine across every single way of fighting. The conclusion draws some points of reference for such a discussion.

The interviews had some considerable cross over because of the differing background of guests, and we didn't stop discussions simply because the conversation wasn't going in a direction we had planned. Overlap was bound to happen – between types of conflict, between strategy and execution, between concepts of fighting, and between views of history. We didn't shy away from these: what we wanted to do was explore some of the boundaries – often self-imposed – to allow those engaged in the profession of arms to focus their understanding in preparation for the future.

The Guests

The interviews covered a variety of aspects on the Western way of war: everything from the tactical level of fighting right to the upper level of decision-making in political circles. Politics, political figures, leaders and economics are covered on other mediums and outlets. Finding political guests who could, and were willing to contribute to the discussion (as opposed to simply regurgitating the standard policy lines that, like very senior officers, they can't deviate from), was difficult. We were fortunate to spend quite a lot of time with very good political animals who had thought deeply about military matters and national security challenges: Madeleine Moon (a veteran UK politician), with experience as chair of the NATO Parliamentary Assembly, contrasted well with James Heappey, the Minister of the Armed Forces in the UK. We interviewed Ben Wallace, the then UK Secretary of State for Defence, which was fascinating in itself given his early and brief experience in the British Army. But it was the retired politicians (and it was similar for military guests too) that provided the real interest – unencumbered by a political agenda and, perhaps, see the national security context with a different clarity. We found that with Michele Flournoy (former US Under Secretary for Defense), David Petraeus (former Director of the CIA), and Jim Mattis (former US Defense Secretary).

This idea of contemporary military theory and the different schools of warfare used to be deeply divided. There used to be a very separate Prussian school of war and a very separate French one and a Spanish one and a British one and a, sort of, Scandinavian one, and an Italian one. Each of them had a very different way of approaching war but they have morphed more and more into a single facet. The single Western way of war that we identified earlier we concluded was the American Way of War that simply everyone else in the West has adopted. The fact that we didn't go to people in Denmark or Germany or Poland or elsewhere for interview, wasn't because they didn't think about it, it was because they couldn't give our audience the breadth that other guests could. While we were highly selective about who we had, we weren't selective in terms of whether they went to a red brick university or whether they made Four Star. It was more about how often they thought and published about war. Was there something in particular that they considered deeply, that they really gave a lot of thought to and when you think about who to interview, they have to stack up against the likes of Sir Hew Strachan, Professor Michael Clark and Professor Beatrice Heuser. These are phenomenal minds and it was hard to think about other Europeans who could make the same level of intellectual intervention of those three alone who articulate their views with such elegance. We did find some exceptional people in the US, Australia, France, the Netherlands, Germany, Chile and in India, but we also hoped that the interviews would be the starting point for conversations, and not a definitive conclusion to them.

Structure

The format of each chapter is based on an episode from each interview: each guest with a particular focus for the discussion, whether it be logistics, urban warfare, politics, media, and so on. As such each chapter introduction provides the reader with – hopefully – sufficient background to follow the discussion and an introduction to the experience of each guest.

Guests would undoubtedly have varying levels of expertise in military matters and in war and conflict. In baselining their views (liberal, hawkish, specialist, tactical, political, commercial, and so on), we took the view that posing an initial question would allow listeners to understand their perspectives about the broader topic in a bit more detail. Over the course of the three series, this question became almost emblematic. The question itself, "What does the Western way of war mean to you?", emerged from a discussion with my colleague and friend Ewan Lawson who provided so much sage advice in starting the show. It struck us, over a couple of large gin and tonics, that the commonality of the question would be a useful access point for guests too. We had no idea that it would become such a useful jumping off point for the show.

Thus, after the introduction to each chapter, a subheading appears, "The Western way of war". This is the starting point to the transcript of each interview. Little has been changed in the transcript: I have added some notes and explanation where required, and removed some of the common linguistic idiosyncrasies of a conversation in order to make it more readable. You can be the judge of whether I succeeded.

Finally, each chapter has a set of reflections after the interview. Many of these have been written some years after the interviews took place. These comments really are a reflection on the conversations that took place and are not intended to update the views to reflect a post February 2022 world (Russia's illegal invasion of Ukraine, increasing Chinese aggression in the Pacific, or Iran's continued stoking of tensions in the Middle East). Readers should understand that the majority of guests on the podcast acknowledged that the world had changed much earlier than mainstream media, governments, international bodies, or some academics represented. To the majority of guests, the world was already more violent and dangerous: the signs were clear. The North Korean nuclear and ballistic missile tests (2006 onwards), the Russian invasion of Georgia (2008), Syrian civil war and use of chemical and biological weapons on the battlefield (2012-2016), Iran's supply of long range munitions to Hamas and Hezbollah (2014), the Russian invasion of Crimea (2014), the increasing sophistication of attacks by the Houthi's against Saudi Arabia (2015), were all evidence that the predictions of Steven Pinker (that war now largely only of historical interest[39], or Noah Yuval Harari (that western populations were more

---

[39] Steven Pinker, *Enlightenment Now: The Case for Reason, Science, Humanism, and Progress* (Penguin Books, 2017).

likely to die from eating a MacDonald's than from a violent attack) were misguided (at best). Those views were, however, largely representative of the political conversations in European capitals that continued until the 2020s. Readers may regard that as disturbing, but the objective of this work is not to highlight the failures of an optimistic political and military system designed to protect and defend. Instead, the aim is to provide context and background for those in the profession of arms such that we might not make the same mistakes again and avoid the attitudes of denial about threats to our values and way of life. The reflections section at the end of each chapter is designed to capture some of the enduring themes from each discussion.

1

Reality Is a Terrible Adversary

In preparing for war, militaries, societies and their leaders have found it difficult to retain a focus on the realities of what combat and warfare entails. Indeed, for a state at peace, simply understanding that war is on the horizon takes considerable imagination. Across history there are few countries or nations that have prepared themselves adequately in time for the onset of hostilities. As the period between conflicts increases, the corporate memory of the costs, sacrifices and skills needed to meet the greatest of all human challenges fades from the consciousness. A narrative often develops about the last war, or even the last battle, being easy compared to contemporary day-to-day challenges in society, and other priorities take centre stage: this might be economic, political, linked to healthcare, alliances, or simply the business of living a busy life distracted from the realities of geopolitical events. Yet within this, there is also usually a societal desire to make sense that the sacrifice that was made during the last war: a storytelling of heroism and daring-do in an effort to justify some of the terrors that will have been experienced. Whether through written words in books, biographies, histories, and fiction, or in films and television, it is common to see the death and destruction downplayed in the narratives and inspirational actions played up (whether individual self-sacrifice or the performance of a military unit). As societies evolve after a conflict, they might not forget, but there seems a tendency across societies to 'move on'. In some circumstances, this might mean taking on a revised identity, seeking a new security arrangement, or moving away from interests of the past, but it usually sees society evolve away from a desire for conflict and contest.

As time moves on, societies evolve further. Long periods of peace distance the memory of combat and sacrifice further from the public and political discourse: that understanding, so common across effected communities in war, recedes in direct proportion to the time since engaged in war. The political and decision-making bodies of a state are not protected from this tendency, perhaps albeit at a slightly slower pace. It is the military which is the slowest to forget, but the ideas, passion, innovation, and understanding of fighting, of *how* to fight, also eventually disappears from these institutions too. This is common across global communities and across the millennia of recorded conflict. Timings may differ but somewhere between 10 and 30 years after a conflict (provided it did actually end and not just become a pause between periods of fighting), any war will eventually have become a moment in history, probably not one viewed as a 'defining' feature of the progression of a society any longer.

When war reappears, whether expected or not, it is a psychological shock for societies and organisations, as well as an economic and, sometimes, a physical one. The intervening periods, depending on length, will have provided societies with the opportunity for change: such evolutions are not just in terms of economic change (perhaps a move in manufacturing across the economy as a whole), but in adopting new methods of production, service provision, and representation. Some of these will have emerged as a function of the previous war, others will be organic, or be transferred from external experiences with other societies. The idea of acceptable behaviours in society also evolve over time, reflected in governments, ruling bodies, institutions, organisations and the military (the latter perhaps a little behind the society they represent). These changes might relate to concepts of identity, norms, behaviours, language, or expectations of the future.

But as societies tend to be inward facing, even sometimes selectively outward facing, they do not tend to spend a good deal of time examining and understanding the wider world. Assumptions are made about how a homogenous world behaves based on those behaviours that have become normative in their own societies, and those that they have the closest relationships with. Those assumptions are based on biases, prepositions, and preconceptions about what is 'right', 'good', and 'valuable'. It is often these ideas that become the founding misunderstanding between states that cause misinterpretations which can lead to the next conflict.

Thus, war comes as a shock: even if not initiating combat operations, warfare in reality requires a sacrifice in blood and treasure that is usually against the norms and expectations of most societies. Militaries are not exempt from this profound experience in the first months of combat. Reality is a terrible adversary.

The key for the successful belligerent is often how you deal with that shock. Which society can be quickest to recognise the changes and react? Which party is able to invent and innovate to match the realities of the character of this war? Which belligerent can make the right decisions at the right moment? Who can recognise the weaknesses of their adversary, and act on them, in a way to provide an advantage? Which leaders can drive a better narrative and convince allies and partners to join (although in many circumstances, just the ability to persuade other parties just to not act against them may be enough)?

This ability to understand an adversary, ideally before war starts but thereafter as rapidly as possible after hostilities commence, is key in maximising the opportunity for success. Divining intent as well as capability is crucial and whilst many promise that Artificial Intelligence will ultimately deliver this, there is little sign that such a system can make allowances for unpredictable or creative decisions that lack precedence, logic or rationality (as defined by the coders of the system). Thus the critical decisions in going to war, and in the conduct of it, will be left to humans for the foreseeable future. Their ability to meet the challenge will be largely based on their own experiences, intellect, and ability to learn, adapt, and overcome. According to General James Mattis, United States Marine Corps (retired), and former US Secretary of Defense, learning from history by reading (a lot) is one of the ways to generate such intellectual deftness.

General James Mattis started his military career in the infantry and served more than 40 years in the US armed forces, ending up as a Combatant Commander for CENTCOM as well as Supreme Allied Commander Transformation in NATO. Beyond that, he was selected and served as the US Secretary of Defence during the rather challenging period of President Donald Trump's time in office. Europeans might describe his approach and, perhaps, his outlook as somewhat hawkish and aggressive by European standards, but it's one that relishes intellectual challenge and informed descent. In his heart, you get the feeling that Jim Mattis is a coalition builder and a big fan of alliances. Now there are more than a few stories, fables, myths and legends about General Mattis, but one that sticks in the mind more than others is Nate Flick's recounting of an incident in his book, *One Bullet Away*, that talks about finding the General down a foxhole with a couple of Marines in 2003. Despite all his other responsibilities, duties and command calls, Call Sign Chaos was living one of his key leadership tenets, caring. The other two are conviction and competence. On leadership, it's worth explaining that Jim Mattis subdivides leadership into three levels: Direct, Executive (which is two-star and above), and Strategic (which is four- star and above).

What Does the Western Way of War Mean to You?

Jim Mattis: I think these kinds of programmes are especially useful right now, as the Western democracies try to sort out what kind of world they want and what kind of world they don't want, what they stand for and, just as importantly, what they stand absolutely against, or what they will not stand for. To me, the Western way of war is encapsulated inside Western values. The Western values here, just for definition reasons, certainly you look at NATO and you see them reflected, but also you can go to Australia, New Zealand, Japan. I mean this is not just the West anymore. It's a western way of thinking about democracy, about people ruling themselves vice authoritarian and totalitarian models that we've seen so many times in the past. So the Western way of war is caught up in society's view of what is the role of war? What should be the primacy of war? Should we turn to it early or late? Should we use it for certain purposes but not for other purposes? Conquering, for example, other people. That definitely is not part of it. So it's mostly a values-oriented framework, within which I look at the Western way of war. Has it been successful? I would only tell you that democracies still stand. Against all the odds, against all the naysayers, the sceptics and the cynics, the democracy of these wonderfully imperfect experiments in governments of the people, by the people, for the people still stand. That brings us to one fundamental point. The roles of militaries are to ensure the survival of our values, of our democratic values, of just and responsive governments.

We have no divine right to victory on the battlefield, so for those who practice this, who are the guardians of values that we hold dear, that grew out of the enlightenment, that grew out of the renaissance period. Today, we carry on with that requirement that we can afford survival, but it's going to take sacrifice on the part of our militaries to ensure, in fact, the Western way of war continues to defend those values.

Peter Roberts: Coming back to those values, would you accept the idea that our values in society have changed? There's no doubt the Western way of war has changed. We went from a reliance on mass and low-tech. We went through airline battle to shock and awe, to the surge. We relied more and more on precision. We increasingly banked on greater technology, we looked at greater information, we became more casualty averse. We liked less risk in many ways. It feels as though those values have changed somewhat. Naturally, they would evolve, right? I mean, I think you're right. You could find those tenets right back in the renaissance, the enlightenment. But they are changing. They do morph. They do evolve with society, right? That feels like we're continuing to see that evolution in how militaries fight today. Would you agree with that?

JM: The military and how they fight evolve all the time. They reflect their society's values. They reflect their leadership. They incorporate and integrate, sometimes successfully, sometimes less so, new technologies, new tactics and that sort of thing. As that dead German fellow Clausewitz said, war is a chameleon and militaries are chameleons. They adapt to their circumstances all the time. Although at least today a fundamental nature of war has not changed, so you find many of the leadership tenets and many of the strategic framing principles remain somewhat unchanged through these changing times. That's because one is the fundamental nature of war. The other is its character. Character is constantly changing, nature less so.

PR: So, on that, I'm fascinated, because I've heard you talk before about being at an inflection point: 'Changing'. Whilst the strategic framing principles, which I think is a really useful term that you just used there, but whilst those principles might be the same, there is this idea that we're at an inflection point. Whether it's because we live in democracies, whether it's because our values are changing, lots of people are talking about this inflection point. More and more, I hear people referring to this because of technology, but it strikes me I've never heard you talk about an inflection point because of technology. Do you think technology plays a core part in driving change, or do you think it's something else?

JM: Well, there are times in the world when the technology seems to be rather stable. If you wanted to call them ships of the line in the 1700s and 1800s or battleships or dreadnoughts in the late 1800s-1900s, we were still talking about basically how do you get big ships with big guns to out-range and out-fire the enemy? Then along came aviation and suddenly technology took it in a totally different direction and the dreadnoughts of the past were basically more targets than they were effectively military instruments. We're always in times when the case of technological change adapts to the larger issues. I would just point out here that for thousands of years, the poets, the philosophers, eventually the lawmakers were in charge of how we integrated things. You integrated them more slowly. Technology changed slowly and look at us today where the folks in Silicon Valley are out in front of the poets, the philosophers, the regulators, the legislators and they're creating capabilities that we could not have dreamed of, or few dreamed of, even a few short years ago. So that's why some people say the inflection point now is sharper and perhaps outpacing society's ability to incorporate these new technologies in ways that maintain, I would just say, support of the human endeavour of life-giving results. In some cases, in many cases, it's a double-edged sword. The technology brings forward something very positive. We can develop vaccines at breakneck speed today and yet we could also create bioweapons at the same speed.

We've got to look at this as a different time. You cannot address the challenges of today using the legislative pace of the past. Now democracies have proven to be very, very agile at times, but so are fascists, so are totalitarians, so are authoritarians. We are going to have to assume that this idea, the values we defend, are going to have to embrace the times in which we live. That means we're going to have to use technology to reinforce what we do. Not to be a straitjacket that we find ourselves tied in knots over.

PR: I always get in trouble for this, because I would always still prefer the poets and philosophers to be ahead of the technology, and people get very angry with me for expressing those views. They want technology to run off. From what you are saying, it doesn't seem that we can hold it back, nor should we, in the competitive environment we're in, right? We just have to accept that either the poets and philosophers get faster, or we have to leave them behind and do some stuff retrospectively.

JM: I think you're hitting the nail squarely on the head. Reality is a terrible adversary. If you take on the reality of technological change at this high speed right now, the technology is going to get even further out in front if you grudgingly go along trudging behind it, saying, 'It's not fair. It's getting out there too far. It's too fast.' So the reality is we're going to have to adapt. Now the thing is, the processes inside democracies can become crustified, cemented in, even when they aren't necessarily fundamental to the will of the people to be responsive. You do have to adapt the processes. Something I've learned over many years with big organisations, sometimes a quarter of a million troops and all, if you have bad processes, and you mix with good people, the bad processes will win nine out of ten. You are going to have to adapt your processes and legislatures and democracies. We're going to have to speed up and integrate more with each other in this globalised world. As the World War Two generation said, coming home from World War Two to America, "It's a crummy world but, like it or not, we're part of it". So, like them, we're going to have to engage more, is the bottom line. We're going to have to come up with solutions that take into account more conflicting interests. It can be done. Look at post-Napoleonic Europe to see something similar in the more political realm, but change is change. There are ways to embrace change but fighting it right now would be futile.

PR: It's interesting because in your background, in the Corps, there is a history of people who embrace radical and fast-paced change: whether it was Brute Krulak or the adoption of manoeuvre wholescale, way before anyone else in other militaries were going for it, there is this idea at its heart that the Corps is always striving to look for the next framework, the next big thing, the next moment that it needs to adapt to. Do you think that that has been inbred right from the start? Right from your time back in ROTC that this is at your core? It's finding this part of adaptability and pushing it forward?

JM: Yes, the Marine Corps could be the most regimented and close-minded force in the world. It doesn't have a service song; it has a hymn. It's been compared to a cult, maybe even a religion, that sort of thing. But a couple of points. Beneath its rather crisp uniforms and its Prussian haircuts, I have found that it tolerates, even celebrates, mavericks more than many of the places I've been since leaving the Marine Corps business - Academia. It's been quite interesting to see it, to look now back at my experience in the Marine Corps for what it was. I think the Naval character of it is one of the things that precludes it from getting too rigid in its thinking. Another thing is you've got a required reading list for all new Marines, Privates and Second Lieutenants. They all have to read a half a dozen books. You start out with mental models and we all need them for how we're going to perform our roles in the world. We may construct them out of movies, or books or biographies. The Marine Corps uses history and biographies. Then when you get promoted to Corporal or Captain, to Sergeant or Major, you get a new reading list. You make General, you go in, you look at those new stars on your collars in the mirror and out comes the hand right through the mirror with another 25 books now you have to read, with many multi-syllable words from Henry Kissinger and all. The point is that you always can look around a table and give Marines a model of what we are going to do or not do, what worked or didn't work on a similar situation in the past. So you're always, to put in field artillery terms, shifting from a known location intellectually.

But the idea too of manoeuvre warfare when it was brought in was that you always focussed more on the adversary than you do on yourself. That does a lot of things to an organisation that unleashes the two qualities you need most. Initiative and aggressiveness. That's what you look for in your young NCOs, your young petty officers, junior officers. Do they have the initiative and the aggressiveness, and then do you train them to practice that judgement so you can delegate authority down? I think that under that condition, institutions get the behaviour they reward. In the Marine Corps, you are rewarded for many things that took doctrine. You had to know the doctrine but you could not fall back as that's the only way to do it. You had to shift from that. We weren't proud of any amateurs that didn't know their doctrine and simply winged it. That wouldn't pass muster. But understanding the difference between a mistake made when you're trying to carry out a Commander's intent and a lack of discipline was critical. I made mistakes at every rank. There was an Army Major who at Fort Sill or Fort Leavenworth who was doing a study of my career and he came to see me at Stanford. I said, "First of all, son, you've got to get out more if that's what your studying, because certainly there's something more interesting than that." He informed me as he went through my career, and he'd done a pretty job. He said, "Are you aware you got in trouble of some kind at every rank except one?" I was kind of irritated that I'd actually missed and how did that happen? But here's the thing. What did the Marine Corps do each time I made those mistakes?

They promoted me. They knew it was my way of learning what right looks like. They weren't lapses in discipline. They were mistakes made trying to push the envelope. I think institutions that reward that behaviour stay out of these intellectual straitjackets that have too often burdened some militaries when they have to confront change.

PR: It's really interesting because when you look at that and you go right back to the start of the Corps and the first World War One engagement, you compare how the Corps behaved in Belleau Wood compared to Pershing or Haig or Pétain, who were rather restrictive in how they thought about it. You could almost feel the manoeuvre approach that was in the Marines at that stage, this idea that 'We will not retreat.' That we will always move forward. That this was us; for very good reasons. The other generals had their reasons for being there but it felt like the USMC were behaving, at that stage, very differently. I think you've said before that, the heart of a rifleman lies at the core of every single Marine.

JM: Absolutely. What you see there is a belief in the Marines that: Number one, your greatest honour is to fight alongside a fellow Marine; Number two, every Marine's a rifleman, no matter if they're flying fighter jets or they're supply guys. But most importantly it's that their attitude is their primary weapons system. So when you are putting that together, you have to look back. Where did that come from? I think a lot of it, again, has to do with naval service. When you get landed on beaches, you either win or you're dead. It's that simple. There is nothing like that to clarify your thinking that; you must be very aggressive, you must move forward. The initiative of every young sailor and Marine in the landing force is absolutely critical. What you're looking at is an institution that has no institutional confusion about its role in the world. That clarity alone brings a co-equal status between a Marine, whether he be a Private first class or he's a General. They see themselves as co-equal in carrying out this ethos.

PR: What's interesting is when you talk about that and then you put it within the context of the alliance structure, where you can identify that *esprit de corps*. You can identify that in parts of the US Army, certainly in some divisions. You can identify parts of that within parts of the British Army or the Royal Navy or wherever else, but it's hard to match that exactly across the variety of elements that you'd find in the coalition. I'm aware when you get up the chain you find that you have very different systems of operating across different militaries. So when you're running a coalition operation, you find that whoever from NATO doesn't want to play this part of the rules, and doesn't believe in that part, and don't have this [capability], it's a fact of life. One of the weaknesses of any alliance is trying to put this together, trying to marry a core ethos and a set of values, if you like, that they can fight together. Yet somehow you manage it. I mean, you experienced that when you were Supreme Allied Commander Transformation. It was one of those big moments, wasn't it, for you?

JM: Well it was but really because you're a Naval service, my First Lieutenant and Captain years in the Marines were spent on six different shipboard deployments, some as long as nine, ten months. You're always in somebody else's country. You learn to do things. When in Rome, do them as the Romans did them. I would bring up that Naval services cannot solve a lot of problems with mass. So often times the Marines find it easier to work alongside allies who don't have all the capabilities of the US Army and the capacity of the US Air Force. In other words, you can only carry so much stuff on the ships. You can never carry everything you want, so you have to make decisions. Everything's got to count. When we move in with some other force, we don't feel like we're some so much larger force. I realise the US Marine Corps is larger than the British Army, so it's a matter of scale but once you're out at sea with the landing forces on a half dozen ships like I've been, you don't have a sense of this is going to be solved simply by adding more to the mix. You're going to have to figure it out and intellectually outclass that adversary. We do remember, we were brought up on this, the three words you must always remember if you want to win on the battlefield. I've had the privilege of fighting many times for my country. Not once did I fight in an all American formation. Those three words are allies, allies, allies. Nations with allies beat nations without them. It's that simple. History is a wonderful teacher here. The role, oftentimes, of US Forces - and certainly of Marines – has been to bring out the manhood [military confidence] in an alliance.

To make people feel confident that this air-ground team that could be brought anywhere by the US Navy could break through to you and be with you if you're in trouble. Think of Kuwait. It [the USMC] can reinforce you. It can cause the enemy to spend an awful lot of time on defence in order to defend a whole lot of places, because it's such a mobile force. I think that one of the responsibilities of our young flag officers and whether they're French or Norwegian or American or British or Italian, is how do you take an alliance where forces have different capabilities and instead of seeing that as a weakness, see that as an asset? You play each to their own strengths. Here's the word I would give to any of the young officers from the democracies: We with malice and forethought are not going to give you the authority that you need and want in order to make this happen on purely authoritative, 'I'm giving you orders to do it.' You are going to have to have a very powerful force of persuasive personality there that allows you to persuade people, that allows you to find the common ground. That allows you to see the strengths of a force that doesn't look like yours, train like yours, talk like yours and use them for their strengths. We're going to expect great things and neither history nor your commanders, your political leadership will accept difficult as the reason for your failure. You're to figure it out, you're to build teamwork and if you think it's all new, take a look at Winston Churchill's study of one of his ancestors there in the fight.

Look at what Wellington faced in terms of trying to weld together allied forces. No sucking your thumb. Just get on with it, figure it out, but you can do it. But you're going to have to look at things with a very un-regimented perspective and play people to their strengths.

PR: I guess the other part about alliance operations and this Naval way of thinking, as you talk about it, is the decision making in coalitions is very different. You've done this at all levels, across your three levels of leadership, direct, executive and strategic. You have covered it across the three levels. It's a very different way of having to make decisions in this collaborative way. A lot of people talk about this as being watered down to the lowest common denominator. To operate at the pace of the slowest. To operate at the level at which you can gain sufficient consensus, but do you buy into that? Or do you think there's a different way of operating and allowing multiple speeds?

JM: It's the trigonometry level of warfare. As Winston Churchill, a guy I'm so proud of quoting because he's smart, yes he put it, 'The only thing,' I think he said, 'harder than fighting with allies is fighting without allies.' If you read the books written about Eisenhower in World War Two, you'd think he spent more time making peace between the Americans, the British, the French, the Polish than he spent fighting Hitler. It's interesting to watch what real leaders are capable of. Look back at Alexander the Great and see how he welded an army together. Look at George Washington. We have this terribly difficult, nasty argument with King George III, and it turns into a revolution. Yet, the whole point of the revolution is we're free men. We don't take orders from anybody. How did he, with French advisors, with Prussian advisors, Polish advisors, British advisors, take an army of free men, guys from South Carolina, who spoke in such a funny way they couldn't understand the guys from Boston? The watermen from Delaware said, 'Thanks very much but I float my own boat. I'm a fisherman. Don't give me orders.' Then the Virginia Grandees who think that, 'We should be in charge of everything,' and find they're not going to be under Washington. How did he do it? He's the most boring leader I've ever studied, but what he does time after time is he listens and he doesn't just listen by not talking. He listens with the willingness to be persuaded, frankly. Then he learns from them how they see the world. That shows respect. He's not into judgements yet.

He's listening, he's learning, he's curious. Then he finds a way to help them. At Valley Forge you might find socks for your troops whose feet are freezing. In other cases, you may just, say, give empathetic reinforcement. 'I know you can do it. I'm confident in you. You will pull this off.' This way, and then he leads. Listen, learn, help, lead. That's the way you put coalitions together. That's the way you find the common ground or create it. That's the way you let people sign up for what they know they can do. For example, I'm in Kandahar airfield and this is right after 9/11 in 2001. It's December and we've been joined there by a dozen other Armed Forces and countries militaries. I get in from my Army Commander in Kuwait a big thick order with some aerial photos and I call in everybody around the table. I throw the aerial photos there in front of them and I don't command any of these folks except the US Marines. I say, 'Okay, there's bad guys up here. They want us to go after them up in the high country, up near the border.' So all the leaders are looking and it looks like a scene out of the bar scene in Star Wars, all these different uniforms. The first guy to speak up was the Australians. No surprise, the SAS say, 'We've been up there. We can do the strategic reconnaissance. We know the ground.' The Marines say, 'Okay, we can lift everybody up to the 12,000-foot level in our big helicopters: we can surround the area so nobody gets out.' The Germans say, 'We haven't been in a fight yet. We want to be there.' So I ask them, 'Are you on our side this time?' 'Yes.' 'Okay, you're on. No problem, you're good to go.' The CIA guys say, 'Our Afghan boys want a ride in a helicopter.' 'Good enough, you're going.' The FBI guys say, 'We want to go up and cut little snippets of hair off people, see if the DNA matches anybody we're looking for.' 'Well, sounds kind of weird but you're on. Bring the HRT [Hostage Rescue Team]. You don't know it but that's who is out there with them. And it went on and on like this, because we had Norwegians and we had Canadians, New Zealanders and Turks and Jordanians and all.

So I said, 'Okay, it's 22:00 now. At 24:00, come back in,' and my Lieutenant Colonel back-up had called the Marine Corps the central coordination element, not the headquarters. I didn't have control or command over any of these forces. You know what those buggers did, Peter? They went back and they used my comms system to call Ottawa and

Berlin and Canberra and, in some cases, got Prime Minister, Chancellor level authority, because they were representing what they knew their force could do. They came back in and with very modifications, we turned out the order that would then put all this in motion, and it worked. It was supposed to be about a nine-hour operation that went for eleven days. It was so rich in targets. You can do this but you've got to have an understanding that other nations take as much a sense of ownership in their troops as the Americans take in theirs. Just get on with it. There's nothing here that Wellington didn't have to deal with, or some of the others in the past.

PR: I guess there's that point at which military officers at a certain level are acutely aware of what their national policy will allow them to do. That leads us to rules of engagement. We're talking about what the Chancellor, what the Prime Minister, what the President of the country is going to allow them to do, how far they can push their permissions. Military leaders now are quite in tune with the politics of their country, which is very Roman, in many ways. It is understanding exactly what the Emperor will go for and what is just pushing too far. But at some stage do you think that because they're so far into that, they forget the realities of what just purely military advice is? I sometimes worry that when you hear people talking about this, they won't even consider options that they feel are not politically acceptable, and yet they're not actually the ones who are responsible for making those political decisions.

JM: Sometimes it could be misused, but that's why best military advice must remain that. Don't try to do your civilian overseers job for them. Give them the options. You will be surprised sometimes that they have freedom to manoeuvre on some things. For example, as one political person told me, 'Don't worry about that. Whether we have 3,000 of you here in this job or we have 1,300, we pay the same political price. So don't get concerned about coming to me by limiting how many troops you tell me what you need.' That would be the first President Bush, who when told how many troops would be needed for how long of a fight to liberate Kuwait, he said, 'I want it done fast and double the number of troops,' basically. We all know what happened in Gulf War One. So I think one of the most important things here is you maintain respect for the others. You don't try to do their job and you remember what your job is. At the same time, if you get into strategic level leadership, you do have to advocate for what you believe is right. The question is, for me, how do you do it without antagonising people? Remember nobody elected we who wear uniforms. The ones who were elected were elected with the idea that we're going to have better healthcare, education, all the things we want for our children's future. So often times, what you're doing at the strategic level in the discussions, is you're reconciling some very grim polarities. That's your job. Don't get frustrated by it but try to avoid antagonism. Best way to do that, bring in historic examples. History's not perfect but it's the best thing we have to light the path against.

Bring in historic examples where other people tried to confront the same type of situation and it worked or didn't, because sometimes that will reduce a sense of personality conflict, and allow you to keep open lines of communication, which are critical.

PR: Let's talk about that civilian military relationship as you saw it: as you went up the chain and you went to two- star, you went to three- star and CoCom. You have a different relationship with the political leaders, the people who were giving you those permissions. There has always been a tension in the US between civilian and military, just as it is here [in the UK]. I remember speaking to a minister relatively recently who was saying, 'Actually, as a civilian going in to sit round with a bunch of military commanders, who are all in uniform, they've all got acres of experiences, they've been doing this for 30 or 40 years, it's quite an intimidating moment. To walk in there and try and get them to do what you want.' That must be quite a difficult moment for them and whilst you get, as you say, the military are getting quite hot under the collar about what they want to do and what they're prevented from doing, actually there's a moment at which the civilian leadership, and you must have seen it in the SecDefs [US Secretary's of Defense] that you worked for when you were CoCom [US Combatant Commanders], they have quite a difficult time too.

JM: It is difficult but it's understandably difficult. The polarities are there and they're not going away. I think they've always been difficult and the more you read of history, I just look at the fight over whether we should land on the coast of France in 1943 or 1944. That was between two militaries and it rose to the political level. This is not pure on the military side if there's some objective reality. Further, there are times when the political leadership is actually correct about how to sustain the fight and the will of the people. Having a good dialogue, and I would say using Hegel's dialectic as a process that allows you to solve problems, knowing full well that simply brings another set of problems, keeps you from being frustrated or going into some kind of non-productive relationship between the principles. Remember too that in the US system, because of that nasty argument we had with the King, that couple of hundred years ago, we actually set up a government that wouldn't work easily with any one branch being in charge. While we have an elected commander in chief, so the military is always under a civilian commander, the legislature has the authority under a constitution to raise armies and to sustain navies. You have to sign a piece of paper when you're nominated for the advice and consent of the Senate for, for example, four-star. The last question on that piece of paper is, 'If asked your personal professional opinion, do you promise to give it to this committee?'. You've actually created a much wider berth of problems for yourself with this, because now you have to bring along a legislature.

Just to really mess it up, we adopted the Romans' idea of a senate and now we have a bicameral legislature that's got to be brought along. It's not easy but in a democracy, go back to our opening remarks together, you're there to defend the values you stand for. If you don't want the job, don't take it. This is part of the terrain.

PR: There is a point that we're getting to now where we can define our adversaries. There are enemies out there who we can clearly define, who are fighting in a very different way than we want them to. In a very different way than we expect them to. In a very different way that we've designed forces to [meet them]. Whether it's Russia or North Korea or Iran or China, they're not fighting the way we want them to. They have a very different way of fighting. In many ways, people I speak to think that we are bound quite a lot by those permissions that our societies impose on us; by how our constitutions are written. This defines how we are supposed to behave on the battlefield, in accordance with a set of values [and behaviours] that our adversaries are not caught in. When the eastern way of war starts fighting the Western way of war, are you optimistic about the result?

JM: Yes, I mean democracies can afford survival. Yes, we try to maintain our values. I remember being told by a Secretary of the Navy, when he was graduating us from our young officer's course, he said, "We are going to need you to do some very evil things. Do not become evil in the process". The fact is that, I think it was the Field Marshall Lord Hackett who said, "You don't win wars by being kind". Now he didn't mean you just go in and kill everybody, but he did say that you've got to be able to do the rough things in order to win. Yes, there's a cost to the human beings that do this. There's a cost to the survivors, not just to the casualties on the battlefield. But democracies have shown the willingness, when the time came, to sustain high casualties. We say that we're averse to high casualties. Well in our Civil War, we were born with a birth defect that we imported from the old world. I know of no other nation in history that fought itself to decisively end slavery. Least of all at the cost we paid. You saw a country that had never taken casualties like this and sustain them year on year. You look at World War Two and the cost to what we call the greatest generation. Hundreds of thousands of our youths, in the Russians case millions, killed to stop fascism. We can adapt. It takes leadership, is one thing that I think we all study in this regard, but whether you're Mandela trying to put a society back together after a hateful war, apartheid, civil war in South Africa, or you're Mannerheim trying to do that twice in his lifetime, after World War One and after World War Two in Finland.

Or you're President Ulysses Grant, not General Grant, trying to put the country back together after our Civil War, there are combinations of leadership portrayed by these people who are thrust forward into these positions that show that democracies can certainly adapt to these kinds of demands, these kinds of challenges and rise to meet them. But it takes a strategic view and, frankly, I think across the Western democracies of 1991, we've been in a pretty much strategy-free mode. The cost of that has been severe. I've been very encouraged recently to see the Integrated Review by the UK followed by a strategy. So often, in the American news, what passes for strategies in the last couple of decades have been simply ways to grab budgets. They weren't real strategy. The British actually decided to budget a few upfront. You could say, 'Well that's not good because it may not give them everything they need'. Well at least they've got a budget they can count on, and now you see the strategic work going on. So it could be that the UK will start leading the Western democracies into a more strategic approach. We've got one, too, in the US, drafted here about four years ago, and it seems to be holding, even during this rather tumultuous change of administrations from Trump to Biden. So we may be getting back to a more strategic approach, Peter, and that's the real answer to your question.

PR: It strikes me that you are rather stoic at the moment [about Western military credibility in the eyes of adversaries] but there is a glimmer of optimism on the horizon?

JM: Yes, absolutely. I think democracies are, again, as Churchill put it, "Trust the Americans to do the right thing once they've exhausted all possible alternatives". I think we've fought wars we didn't need to fight.

We can get a strategy out there that is not just bonding allies together, it's also an appetite suppressant that you don't go off and just try to be all things to everybody. It actual limits you to what are your vital interests. There I think the strengthening of NATO under the current administration is a very positive sign. The strengthening in terms of the US is with you. There is no equivocation.

Reflections

General Jim Mattis is often referred to as a soldier scholar: to soldiers this means someone who carried a veritable library of books with them whenever they deployed (in addition to their significant home library). To a scholar, the term does not mean someone who reads a lot: rather, it expresses a set of experiences that the person can relate across complex problems, and inter-relate that analysis with leading-edge literature. A soldier scholar is, simply, a thinking service person: someone grounded in experiences and knowledge that is more than their own, but equally brings something unique to dialogue and discussion based on military service. It is, perhaps, that sense of history and self-reflection that comes across when you meet Jim Mattis: full of the honest and thoughtful words of others. This kind of scholarship also keeps senior officers free from hubris – something all too quick to slip into as an accompaniment to the trappings of high military rank and/or office. Mattis' ability to not only prioritise his own time for self-study has always been accompanied by periods of reflection on what he has read and how it applies to his own context – whether of the fight he is in, or the style he needed to adopt. His own experiences in uniform, whether during the Cold War, in the full spectrum of conflicts he served in, or the military and political positions of command he occupied, necessarily shaped his view of the world and the national security challenges being faced.

Yet despite all of that baggage, Mattis was able to elevate himself to higher almost philosophical questions of democracy, values, ethics, and values. In doing so, I did not find someone who simply regurgitated meaningless phrases (like 'rules-based order'), instead, it was clear – to me at least – that the US military has an ability to generate, promote and reward officers who think, who diverge, and who digress from the norms.

In thinking about the Western way of war, Mattis' focus was less on the ability to fight a certain way or with a homogenous doctrine, but rather was clear about the continued relevance of people, of creativity, and of the enemy. The idea that an adversary has as much impact on a battle as your own forces is sometimes forgotten – especially during long periods of peace: there is no divine right to victory for Western states or their militaries. Indeed, per Mattis, reality is a terrible adversary.

2

Operation Tethered Goat

Challenging the practices, process and doctrine of Western militaries is something few are able – or willing – to do. There are, occasionally, internal debates within confined military circles that seek to unpick the assumptions and presumptions of a force design and the preconceptions upon which it has been built. Yet there is little evidence that any of these discussions have had a significant impact in terms of evolving a *way* in which militaries fight, let alone the self-reinforcing conversations between the military and their political leaders. This might be the reason why the Western (American) Way of War has not experienced any radical change since the 1980s. From 'shock and awe' to counterinsurgency, Western militaries have returned at every conflict to an increasingly refined stand-off form of war: one that depends on intense surveillance, connectivity, air power, fighting at range, precision, and boutique weaponry. However, there is an issue with this idea of military superiority through airpower, information and distant control of events: it may deliver reduced casualties on battlefields, but it has yet to win a military campaign. The assumption that a theory of war based on technology, money and a minimal risk tolerance can deliver a preordained right to victory lacks an evidence base.

In seeking to revolutionise the Western way of war in 2016, US Deputy Secretary of Defense Bob Work sought to re-establish a US competitive advantage in conventional and deterrence capabilities through the US Department of Defense's Third Offset Strategy: a doctrine to rebuild America's conventional force overmatch in order to deter Russia and China. Started in 2012 by US Defense Secretary Ash Carter's Strategic Capabilities Office, the idea had sound beginnings. After examining the emerging military investments by Russia and China, it was noted that potential adversaries were building military tools capable of challenging and overcoming the US' military advantages in many domains (space, cyber, command and control, mass, sea power, nuclear weapons, data, air superiority and connectivity). To counter this, a distinct choice emerged for a next generation of US defence policy: either recommence a rapid and massive US recapitalisation and an increase in size of the current force, or plan on a cheaper alternative that hunts down exquisite capabilities able to unpick an adversary's asymmetric advancements. The idea for a Third Offset was based on studies from the 1970s and 1980s over the Revolution in Military Affairs (RMA), a popular concept in NATO and with Russia that – it was hoped – would provide a disproportionate military advantage without the burden of eye-watering and enduring costs. Core to the concept itself was a belief that government spending alone could provide a sufficient stimulus for industry to innovate-to-order for the Department of Defense (DOD), and deliver the efficiencies of business without the costs of development associated with internally developed solutions by government agencies. What was not to like?

Yet the Third Offset Strategy weighed the future gains as greater than investment in the current force, leading to a weakening of standing US forces. In trying to get the doctrine off the ground, the US invested time and effort in anything and everything that might have the potential to deliver an advantage. But there was no consideration within it as to how to employ or cohere this random set of new capabilities. Despite this investment of time by DOD leadership, there was a marked absence in a unifying theory of how the US might fight in the future, armed (as it would be) with random capabilities developed and procured not through a demand, but simply because they existed.

So, despite a significant investment of cash and intellect, the West remained entrenched in a form of stand-off warfare, ideally suited for wars against non-state militaries, or even sub-peer state adversaries who adhered to the idealised form of operations that US and Western forces were, in theory, well matched to counter. Increasingly, however, the democratisation of complex and sophisticated weapons to proxy groups by state actors made sub-peer militaries more capable of challenging well established and technologically capable military forces. The Houthis in Yemen were able to conduct sophisticated strikes on Saudi critical infrastructure using a combination of space imaging to target key nodes in oil production pipelines with drones, cruise and ballistic missiles penetrating deep behind the established front lines of battle. And the Houthi's did so without a reliance on air power, basing rights, complex logistics infrastructure, or detailed and beautifully connected command and control networks.

Adversaries were no longer a tethered goat waiting to be slaughtered by a smart bomb delivered by a stealth jet from over-the-horizon.

Given this evolution in the way adversaries were choosing to fight and engage in contests, there was – and remains – an increasing chance that a first-rate military force would not only lose to a peer competitor (Russia or China), but that defeat by a sub-peer group was also on the cards. But there was still no realisation that the Western way of war represented the way Western militaries wanted to fight, nor the way they might need to fight in order to win.

It is here that the role of external intellectual challenge to the norms and narratives of Western military thinking becomes essential in thinking about what comes next. Breaking the groupthink of military personnel, mantras and doctrine is essential if shocks and defeat are to be avoided.

Wilf Owen is one such challenger. A former soldier turned scholar, Wilf's adversarial style challenges many in the military community. He has been passionate about how the West needs to, "Fight with Americans, but not like Americans". The idea that a single Western way of war may not be suitable or working for every military might be politically and militarily unconscionable to the resident epistemic community of leaders, but differing forms of war evolving as a complementary military approach has historically been a much more successful approach for allies over history. It broadens the attack surface and makes allies less predictable to common adversaries.

William F. Owen, known as Wilf, is the co-founder and editor of *Military Strategy* magazine, formerly known as the *Infinity Journal*, which has been in publication for over a decade. He has also consulted for the British Army and others on a range of military, command, doctrine and capability issues since about 2003. He served in the British Army, both regular and reserve, for just over a decade, before working on defence projects in West Africa and the Far East. He holds a master of research from the Cranfield defence and security part of the Defence Academy of the United Kingdom. Wilf has published extensively on matters of warfare and fighting, reflecting the views of much of the country in which he now resides, that of Israel.

What Does the Western Way of War Mean to You?

Wilf Owen: Well, to me personally I think it has differed or it has evolved from what the phrase originally might have meant, in that it descended from Liddell Hart's famous A British Way in Warfare. The Western way of warfare is clearly and can only be how the West has chosen to conduct military operations for maybe the last 25 or 30 years. And it's characterised by, at the high end of warfare (taking on nation-states or large armed groups), by large amounts of air power, air superiority and everything basically flows back down from there. It is a stand-off form of warfare, which isn't necessarily wrong, but it can be very easily challenged. I think the more sensible opponents in terms of the non-nation-states are probably already – or have for quite some time – arguably been looking to dislocate that as being a go-to approach.

Peter Roberts: It's interesting that you focused on air power: To me, what you hit there were the concepts of power and superiority – superiority being one of those key words that I think flows out of quite a lot of military leaders, military doctrine, military thinking in the West. I'm using this, amorphous mass of the West meaning most states, and the exception to that being Israel, who probably fights in a slightly different way. But it seems this way of sort of addressing adversaries as if you have this superiority has not just been applied in the high end, which it was designed for as you raise, but also across a number of other types of engagement. It's similarly been applied against ISIS; it's been applied in 2003 in the Iraq war; it's been applied in counterinsurgency. It's the same set of principles that have been applied that across the way that NATO and American forces have fought for the last 25 years, right?

WO: I would say it's process driven. It's not necessarily a choice. If you want to make close combat anti-climactic – in other words, you don't want to risk people in close-range fire fights – you're looking basically to win the battle by use of, in very, very generic terms, use of bombs and artillery. Now, bombs and artillery have become fantastically more precise, meaning you now use them on a room-by-room or even a seat-in-the-vehicle basis. Most of your sensors are in the air, so if you can connect those means of fire with those sensors, it can have a very, very decisive effect. A really good example of this is the recent use of drones and indirect fires by Türkiye in Syria, which showed them going out and just literally systematically servicing targets. Now, as a way of war, this works and it's coherent, providing you're up against an enemy that conforms to that model. Clearly, problems arise when the enemy can deny you those means of operating.

PR: So just to expand on that a little further, when you say 'can deny it to you' you don't mean that they can deny you the targets, right, or is that what you mean?

WO: They can do both. They can deny you the targets in terms of, you know, they will do the classical, tired old 'they will move into urban terrain and they will be amongst the civilian population,' which is a well–chewed-on aspect, but more to the point, that if you look at the supposed what we choose to call peer-or-near-peer players, they will just start knocking down your sensors or they will go after your networks electronically or kinetically by shooting down your unmanned aerial systems which are selecting the targets or even shooting back at your fires platforms. That's why as a way of war this works really well against the second or third eleven. It doesn't work against the enemy's first eleven, or an enemy that's capable doing the same things.

[The references to 11 here is a cricketing metaphor: there are 11 people in a cricket side, the best players being in the first eleven, the lesser players being in the second or third elevens].

PR: So it effectively needs to change. Do you subscribe to the view that even second-tier adversaries now have a level of sophistication, weaponry, reach and speed, as well as a way of operating, that now makes them almost first-tier enemies?

WO: Yes. Take anti-shipping missiles: we've seen with the Houthis in Yemen with weapons like that, where getting hold of comparatively sophisticated platforms, which are relatively easy to hide and relatively easy to employ, means that you're chasing their firing platforms and they're not necessarily chasing yours, because they have a greater diversity of targets that deliver their policy aim. I mean the West is, should be, limited to using armed force against armed force. A lot of the people you're going up against are not necessarily constrained in that way, and can use armed force against, for want of a better word, high-profile targets that may or may not be military in nature. So, yes, it's the classic problem of dislocation and people going after different objectives with different means.

PR: If the adversary has adapted the way they're fighting and there has been relative stasis in the way the West is fighting – there might have been evolutions and improvements and innovations within it, but the core system of the Western way of warfare has remained the same for this 25, 30 years – does it now need to change? Are there sufficient drivers that it now needs to accept that something fundamental needs to shift about the way the West fights?

WO: Yes, I think there are. The spectres of Russia and China, as the two names we never seem to mention in terms of peer competitors, are absolutely it. Whether – if an enemy has large amounts of ground forces and can deny you the use of your electronic spectrum and can deny you the use of your aircraft and unmanned aerial systems and attack helicopters, then they've presented you with a significant challenge and I think the important thing to understand here is the West, if we're calling this a Western way of warfare, its development is accidental. I don't think anybody actually sat down and planned this, because how we planned to fight the Cold War was AirLand Battle and we took account of the fact that the enemy had sophisticated air defence systems and things like that. We've fallen into this because a lot of those systems stayed around a very long time in terms of capability, and they were very easy then to use against an enemy that couldn't contest them. So I think it is time for a fundamental rethink about how we, well if nothing else, protect our ability to use those domains. That's a fairly sophisticated technology-based conversation that has to flow from better ideas rather than just better technology.

PR: But the US would argue, in places like US Army Futures Command and those in the Pentagon, that the Third Offset Strategy, the new National Defense Strategy, the idea of dispersion, these were new ways of warfare for the Americans. Do you accept that?

WO: If we're going up against somebody – let's be honest, the really frightening thing is losing militarily to someone who is an actual peer competitor or – and the term 'peer' is debatable in terms of its utility – but if you suddenly had to go up against Russia or China, there might be areas where they could easily over-match us. I say 'might be' because until you have an actual real war you don't really know. I mean, a lot of this is guess work and a lot of this resides in the classified domain and probably isn't up for public discussion. But in terms of inoculating ourselves against Russian massed fires, we absolutely need to disperse. We need to be less reliant on the received wisdom of Cold War. This is, it's a funny thing because people say what if you're wrong. My argument with this is even if we're wrong, we're still doing good things. It's always good to disperse, it's always good to reduce your reliance on the electronic spectrum. It's always good to reduce your reliance on things that the enemy might find easy to contest. So, yes, I think that conversation is long overdue and I don't think there's a lot of risk in having it.

PR: I worry that this new US way of fighting – and they're sort of saying it is – isn't really a new way of fighting. It still relies on precision, on range, on speed, on strike, on minimising causalities, on air power, on superiority, on sort of stand-off warfare. It seems to be very much an iteration of what you described as the Western way of warfare 25 years ago. It doesn't seem to have evolved. It might use new technology, but it hasn't changed that much. The kill chain is still the same. The human-in/on-the-loop is still the same. There doesn't seem to have a fundamental conversation about how Western military forces will need to fight, given what you say about peer competitors.

WO: I absolutely agree with you and I think that one of the real problems is if we look at who is involved in the conversation: for a start, not a lot of people are involved in this discussion. If you look at the vast majority of output, of think tanks, academic papers, the entire industry, very, very few people are concerned about methods of fighting and operating in specific terms. There's some quite abstract on-the-one-hand this, on-the-other-hand that, need for more research kind of approaches. But the actual number of people who can talk in detail about how to fight land warfare at the applications level is very, very small indeed, and it's mostly Americans. I am, as you know, renowned for being less than convinced that fighting like the Americans is a good thing. As one British general said, we will fight with the Americans, not like the Americans. Because of the peculiarities of the British defence budget – it's nothing to do with sovereignty; it's all to do with money – we have to find better ways of using force that don't make us reliant on the Americans, because the American approach to warfare is probably not affordable for the UK in the near future. So we do have to find better ways and different ways of working.

PR: If the drivers for change in the Western way of warfare or how we fight – technology, money and the adversary – if those are the three key factors that you think are there, where do you think might be a good place to start looking at this? Are there other models? If we looked at the Israel Defence Force (IDF), for example, do we think the Israelis have some original thinking, and there were some thinkers – Shimon Naveh, and others, who came up with rather different ways of looking at the problem, but perhaps not different ways of fighting.

WO: There's always a problem with Israel, and thinking with my Israeli hat on, is that I think Israel is good to examine in understanding how they do things. I don't think Israeli wisdom travels. In other words, I don't think the Israeli approach necessarily transfers outside of Israel because Israel has developed very, very specialised techniques to addressing particular policy and military aims. I mean, the driveshaft between what the IDF do and the cabinet and high-level decision making is almost direct, with almost no interfering gears. They have a very, very specialised system that works very well for what their policy objectives are. The real lesson, I think, is look at the size of the Israeli defence budget. I mean, regardless of US foreign military assistance, look at the actual size of the budget and then look at the range of utility they get out of that, including naval, air and particularly their land forces are very large compared to their level of investment. It literally is bang for buck. The thing missing from the UK conversation is a realistic discussion on budget. In the modern (whatever) way of warfare that Great Britain has adopted, money is absolutely critical, and the money discussion simply hasn't been had because people tend to focus on things like how many infantry brigades can we fund. Infantry brigades might not be the answer. Aircraft carriers – we've got aircraft carriers...

PR: But no aircraft to go on them, for example…

WO: Yes, an aircraft carrier without aircraft, could – I'm not a naval expert, but it always strikes me as an intriguing challenge, yes. Buying the aircraft carrier is only, I think, 25 per cent of the problem. You then have to fund the rest of it. It's absolutely the same with land forces. There is no point in funding four infantry armoured or strike brigades or any form of brigade you have unless you can afford the 90 days a year in the field that you require to train to the level where you can effectively use these things. I'm picking 90 days a year as the old BAOR [British Army on the Rhine] figure. It might be with modern simulation or whatever we don't need the exact same amount of time. But the fact is that buying the capability without investing in the means and methods to train for it is clearly negligent and that needs to be a conversation you have to have if you're going to develop a way of fighting that you cannot only afford to procure, but you can also afford to train with. Israel's only edge over all its neighbours, historically, has been training. People can harp on about technology, but you only have to look at [the IDF in] 2006 and see how rapidly things go downhill if you put training in the backseat and say we can do everything by using the air force and a bit of cyber and a bit of information. Land forces need to be well trained, and that's the takeaway lesson from 2006.

PR: Can I just go down a rabbit hole here for a minute? We'll come out in a minute, but I want to talk just for a second about training with you. There is a huge move to reduce the cost of training by putting it all online and in a virtual world and a virtual globe atmosphere with virtual adversaries and allow people to fight in sort of X-box style. What have you seen from this? Is it effective? I was reading something the other day that said that no matter the amount of online training that you do, you cannot replicate the force pressure change from when you fire a round, whether it's on a ship from 155 mm self-propelled Howitzer, from a rifle, the change in air pressure is around you, it's the sound that is the one thing that changes. You cannot replicate that feeling and the sense of combat: giving people the edge, to me, seems something quite different from what we think virtual training can deliver.

WO: I would actually be more fundamental about that. In my view, and mine may not be a fantastic military experience on which to base this, I would say the one thing that simulation can never match is changing a track at night, in the pouring rain, when you haven't slept for 36 hours. You can't simulate changing the power pack on an armoured personnel carrier or a main battle tank. You can't simulate trying to give orders in NBC kit when you haven't eaten for three days, and that – armies that spend lots and lots of time in the field are basically inoculating themselves against the basic human hardships of operating and being able to continually operate. Just being able to deal with cold, keep yourself dry – a lot of the sort of technical aspects of combat are processes and procedures but what makes those processes and procedures work is having the manpower with the confidence, the training and the physical endurance to actually make them work as and when it should. The example I always give is to imagine marching 200 kilometres and then conducting a river crossing to move straight into a forming-up point, to then cross the start line into an attack, all done on radio silence. You cannot replicate that in the virtual environment. It's something you have to go out and do in Northern Europe in the winter, or in the Middle East in the height of summer, because —

PR: And hopefully more than once every five years….

WO: The figure I pick is 90 days a year, and everyone always says why do you say 90 days a year? To me 90 days a year is the start point that you've got to do, and that means at least four major field exercises a year. If various iterations of training prove that's too much or not enough, or whatever, but you have to have a start point, and certainly the received wisdom from people who trained in field army back in the days of BAOR (which was the last time we took that level of military engagement seriously), that seemed to be the figure. Lots of things, like gunnery, you can do in a simulated environment. I'm sure the technicalities – you would know better than I would – the technicalities of ship driving can be learned in simulation to save time and money, the same way as the airline industry do. Likewise, tank gunnery can be perfected in a simulator and then confirmed for real. But the idea that you can do unit level training online and that you then don't need to confirm it in the real world is, frankly, an ill-founded conjecture, to put it mildly.

PR: So now we can zoom out. What you hinted at slightly earlier was this idea that we're almost never going to have the right force design for the next fight. Standfast the Israel example, which pretty much knows where it's going to fight and therefore probably how it's going to fight, it's got enough information about its adversary. For those powers that don't know, you seem like you're in favour of a spread of capability and then focusing on the one that's most appropriate for the fight that you go to. Would that summarise what you think? It's about adaptability.

WO: I've a pat phrase, which is you never know where, when, why or against whom you're going to fight. Mind you, it's fallacious to actually predicate any of your ideas on having an exact solution. Just to take an example, the idea that you are definitely going to fight in the Baltics, for instance, is going to lead down a Baltic-centric rabbit hole, which could find you wrongfooted if you fight somewhere else. The thing you have to be absolutely certain about is how you are going to fight, regardless of whom you are fighting and that is a skinnable cat. That is not something that's abstract or impossible to assess, and that is where training comes in, which is if you train overwhelmingly to address the basics, and there aren't new basics. You always hear these sorts of online disruptors saying, ah yes, but there are now new basics. Well marksmanship, maintenance, logistics, administration, keeping warm and dry, begin able to plan quickly, being able to deploy quickly – I'm not talking about bootlace level stuff; I'm talking about higher formation. Higher formation and joint force fundamentals will adhere – a really good example, the RAF, is the way you can quickly load, quickly and safely, load transport aircraft is always going to leave you in good stead. In the Royal Navy, the way you can quickly and safely continuously operate helicopters off small ships is always going to put you in good stead. Again, it's the way you can keep operating a ship in very bad weather is always going to make you a better navy than other people. But again, a lot of this stuff seems to be absent, certainly in my limited view. It seems to be absent from the grown-up conversation because ultimately if you want to have a Royal Navy that can operate in appalling weather, the thing that puts that ship in appalling weather is money, because it's no good being in appalling weather once every five years. You as a naval man may tell me I'm wrong on that, but I think that's a reasonably good example. The same with the army. There's no point sending the army out into a wonderful, lush, British countryside. You want to send them somewhere where the weather is absolutely appalling and you need a high level of personal administration to keep yourself going. If you're better at all of those things than your competitor, then those are all good things which you can safely train for, and you can train for within a budget. So again, to me, it shouldn't be a mystery, and I'm a little bit surprised that it's always presented as a mystery because ultimately you will never be able to predict when, where, why or against whom that the thing, the only thing you can focus on is how, and how is a skinnable cat.

PR: You said it was unlikely that everyone would be able to keep up with the pace of US technological change, that it just was impossible for powers with smaller budgets, but that they wanted to keep the same broad spectrum of capabilities in terms of platforms and people. The one thing they need to do so is to harmonise their baseline training up to echelon level, the highest formation level possible, which gives them interoperability, and then at some stage when they are entering a fight, they need then to figure out which part of the force is going to be most applicable and focus on deploying and using that. Is that what you're saying?

WO: If you're looking at what is essentially a modular approach, you have to train to be modular. You can't just tick the box to say 'modular'. You actually have to do the training to say that – we can't just throw it together at the last minute. We have to know how to throw it together at the last minute. The classic example, perhaps, is the Falklands, where just good solid military practice won through. It wasn't perfectly executed, and it wasn't without challenges, but the Royal Navy, the British Army and the RAF were probably one of the few organisations on the planet in 1982 which could have done that with the degree of limited infrastructure which they had to go away and do it. I tend to believe that – it's not a technology thing. When I say you can't keep up with the US, what I mean is you can't generate armoured divisions, you can't generate brigade combat teams like the US. You are going to have to pick your fight commensurate with your budget and then you're going to have to be able to bring that thing to the party, which the other side finds hard to compete with. In terms of the UK, it's all going to be about interoperability with partners and allies. The sovereign bits that the UK always has to be able do everything alone, I'm not saying it's wrong, I'm saying it's challengeable, and it could well be that the UK wants to get out of particular areas of competition. I mean, particularly the idea of armour; the obvious one that I've talked about at RUSI is war without tanks. I'm a fan of tanks, I think tanks are great. In my spare time I go and look at tanks, I'm a sad tank man. But the one thing it comes back to is these things cost a lot of money and they're only relevant to a certain fight. We can all make jokes about dinner jackets and that, yes, but when you really need a tank, you really need a tank. That's a challengeable view. It's not certain. There's no evidence to say that it's absolutely correct. Yes, tanks make things easier. But tanks are heavy and complex and once you involve them in a formation you are placing logistical and engineering and equipment support burdens on that formation, which would not be there if the tank was not there. I'm not saying get rid of tanks tomorrow but I'm saying a lot of the received wisdom of land warfare, which has occurred since the Second World War, is challengeable in terms of having a UK army that is now based in the UK but might have to go anywhere in the world to operate and fight. If you want to get there with speed and sustain yourself there in a timely fashion with low levels of logistic robustness, for want of a better word – in other words, with lines of supply that might be compromised either physically by enemies or electronically – then you need to have systems that will allow you to do that. Again, that's the conversation I'm not really seeing – and I'm sure smart men in the Ministry of Defence are having it – but it's not out there in the public domain. That's certainly true.

PR: It seems very few people are having it. Whilst I agree with most of what you said, I sort of take a little bit of an exception with the example of the Falklands War. I think the global distance thing about the UK being one of the few states that could have done it back in 1982 is fine, but we're talking about almost the perfect engagement for a Western military at that stage. There were all the things that we expected. There was anti-submarine warfare, there was anti-surface warfare, there was anti-air warfare, there was a bit of strike, there was a bit of light infantry fighting. There was the use of radio equipment and data links. Apart from the supply chain, there was nothing unexpected in that fight at all. It was, in fact, the perfect the engagement for the British military at that time. For me, it almost reinforced some of those lessons that I wish we'd unlearned at that stage. Sure, it proved the need for global deployability, but apart from that I find it a less compelling example of Western superiority or Western way of fighting because it was absolutely what the British military had trained for the previous 40 years.

WO: I'm not actually disagreeing with any of that. My point is the training the British military had done – let's be honest, the Argentineans went and set up on an island, with conscripts, and then it was Operation Tethered Goat from pretty much then on in. I mean hard and demanding fighting? No it wasn't. It was certainly no cakewalk, and there's some well-earned VCs and DSOs along the way. But the fact is the British military were only in a position to do that because of the training and ideas that were in place at the time. Could the British have fought it differently? There's lots of military debates, and yes in a certain way it did reinforce [contemporary thinking], but my point here is if you want to do things like that, and you could pick an entirely different example, unless you actually have the fundamentals in place, in other words the peculiarity of the Falklands War wasn't anything to do with British equipment or defence policy. It was the fact that British military training and a doctrinal approach had laid the groundwork where, whilst it was demanding, it was nothing that was deemed particularly impossible for the British military to do. I think the Falklands is perhaps a bad example on the technical point, but it would certainly be possible – at least I hope it would be possible – for the militaries of today, the armed forces of today to turn around and say, okay, so what do we think are the equivalent fundamentals that we need to focus on today. If I could only suggest two or three, which would be the ability to truly rapidly deploy, to be able to strap together an effective force from the components you have and to be confident in doing it, so that's two things. To be able to manufacture the force to the point where it can rapidly deploy, and when you get there to be able to sustain it so as it can use the levels of lethality it needs to provide, for want of a better way, political utility, because lethality in and of itself, unless it's precise, proportionate and discriminating, has no political utility in the modern world. Lots and lots of 155 shells might not be the answer. In other words, that's not necessarily the debate we want to have, but what is useful politically in terms of military strength might be a very useful debate.

PR: I want to pose a final question because I think you might have an interesting take on it. The evolutions in the Western way of warfare hopefully will happen. Fundamental rethinks or not, there will be evolutions. But they won't just be our evolutions. There will also be those of the adversary. Do you think that the adversary would react more swiftly to our changes than we could change, particularly when we do so much in an unclassified open way these days, or do you think that somehow we'll be able to come up with a new way of fighting that completely discombobulates the adversary in ways that, in many ways the idea of precision, fixed the Soviets in 1982?

WO: I think it's entirely possible to come up with a new way of fighting, but perhaps you don't have to. Ultimately you have to develop a way of operating around which it's achievable in terms of procuring equipment and training and being able to sustain it. You don't have to come up with one way. What you've got to come up with is a general approach. I think a general approach is very difficult for anybody to counter specifically. This is a competition. At the end of the day, you do get down to a boxing match where just being better, stronger, faster and more aggressive will reap dividends. There isn't some domain of warfare where we can seek and achieve unique superiority for the foreseeable future. That just doesn't exist, especially in a world, as you say, where it's massively interconnected and everybody has a cell phone camera. I don't see any of these things as necessarily being the game changers that some people suggest they are. They're just another thing you have to deal with and another thing you've got to cope with. I mean, I think the point where we would get beyond our current adversary, where we would make life difficult for them is if we really started to question our own approaches to really work out what does make life difficult for them, because really my whole thesis is that's what's been missing from the current debate. We are still having a British defence debate which is, in many ways, still fixed – certainly on the equipment side, and that's all people seem to talk about, is equipment. But if you think that there's a debate in the land warfare domain of saying armoured infantry brigades, numbers of infantry regiments and metrics like that, then we're very quickly going to default back to a BAOR [Cold War] model of fighting and operating, which is completely different from that which we need to have in terms of being a force that provides UK, NATO and whoever else with real political utility. I don't even think it's a matter of coming up with something our adversaries find hard to compete with. We just need to come up with a way that is affordable and doable. That's what they will find hard to compete with – is something that we can afford to do, because at the moment essentially adversaries could just price us out of the competition. More than anything else this is about money or the lack of it, and a post Covid-19 UK defence budget, I think, is going to be a very interesting world indeed.

Reflections

Wilf Owen always challenged me about the language of war versus warfare: one being a political paradox, the other being a military reality of combat operations. Whilst the Western way of war could be perceived as both of these (and many of the conversations about the topic move between the two), Wilf's remarks remind us not to allow clever or imprecise language to escape the realities of what politicians will be asking of the military: the fighting part of war and warfare. As such, the focus of the discussion was based around a reality of military capability: not simply a spreadsheet of tanks, ships and aircraft, but the less quantifiable elements of martial power. Training, education, grit, aggression, lethality, survival, logistics, coherence, control, education, creativity, determination and so on. Simply having an armoured division or a couple of aircraft carriers is not enough to deter, coerce or defeat an adversary in the contemporary world of national security.

It is from these elements that the credibility of a state military is truly derived. As evidence in numerous contemporary conflicts has found – and has been particularly clear in Ukraine since 2022 – the equipment a military fights with has become less important in their ability to fight successfully than the ingenuity they employ in undertaking combat operations. This should not be a surprise to those even vaguely aware of the long history of humanity's various wars. Yet these facets do not seem to be part of the discussion about war and martial power in Western capitals, nor prioritised in funding options. As Wilf pointed out, the predilection for political and military chiefs is to engage simply in rhetoric about technology, fiscal budgets, and gaining some kind of asymmetric advantage. It is disappointing that this has not changed (for the majority of Western states) since 2022.

3

People as the Decisive Edge

An examination of conflicts, whether contemporary wars or those from different periods of history, is instructive in trying to determine what becomes important, even decisive in combat and conflict. Evidence, rather than passion, emotion, conjecture, and the fictional stories of techno-fictionalists, is key to separating reality from mythology. In divining the essence of success in warfare, it is also useful to determine what is revealed from a scientific examination.

When analysing the critical factors to success in warfare, the foundational principle that militaries use is termed centre of gravity analysis. This is a process that seeks to divine the single more important factor around which all things gravitate (whether in a plan at the strategic, operational or tactical level, as a problem set that can be unpicked through it, or the analysis of an enemy's vulnerabilities): in the original text it was defined by the originator of the term, Carl von Clausewitz, as, "the hub of all power and movement."[1] In contemporary military language the US Department of Defense defines it as, "The source of power that provides moral or physical strength, freedom of action, or will to act."[2]

It is peculiar that militaries understand themselves to be complex organisations yet seem to believe that a huge number of variable factors can be distilled into a single factor. Yet there is something to the identification of some of the critical variables in war that must be undertaken – to build

---

[1] For a longer and fulsome examination of centres of gravity, see Joseph Strange and Richard Iron, 'Center of Gravity: What Clausewitz Really Meant', in *Joint Force Quarterly*, no. 35, 20-27. https://apps.dtic.mil/sti/pdfs/ADA520980.pdf.
[2] US Department of Defense, *DoD Dictionary of Military and Associated Terms*, US DoD, November 2021, 30. https://irp.fas.org/doddir/dod/dictionary.pdf.

on one's own force advantages, and to undermine an adversary at the most decisive point.

For many Western militaries today – and there are quite different definitions of a 'centre of gravity' even between NATO states – the oft used claim is that their greatest strength is built around technology. It has become a popular mythology and soundbite for political and military leaders that the belligerent with the technological superiority is preordained to be victorious. There is little evidence for this in history. Indeed, in trying to divine a centre of gravity, it is not technology that provides Allied forces with their greatest source of strength, freedom of action or will to act. Arguing this to a contemporary military audience is often, however, futile.

Instead of attempting to push water uphill, it is worth trying a different path: looking at fundamental rather than strategic capabilities. The former can only exist when it fills the following three requirements: (1) it is valuable, (2) It is rare, and (3) it is immutable (for example, that it is difficult – or perhaps impossible – to copy). Taken in this way, technology (certainly a critical and indeed strategic capability), cannot be considered to be a fundamental capability; it is certainly valuable, and quite possibly rare too. But technology is not immutable. No single state or individual has the sole rights to weapons of war, whatever they are. Hypersonic weapons, ballistic missiles, global precision navigation systems, cyber payloads, 6th generation fighter aircraft can all be developed by different entities. They are replicable – and the good designs often are. Submarines are certainly valuable and rare, yet they have been copied and are now widely used by various states around the world. This analysis does not mean that they are unimportant: submarines might well be judged to be platforms with capabilities that have strategic importance. But they do not represent a fundamental capability.

Technical skills or capabilities of any kind fail to meet the requirements of a fundamental capability because they are always copyable. Those things that do fill the requirement set of the definition are, in essence, human – precisely because they are so difficult to replicate, but they are also rare and valuable. These facets might be the leadership abilities of the NCO cadre, the brilliance of certain generals, the intransigence of military units when faced with impossible odds, or the political will to fight of a nation state. All of these rely on the previous history and culture of an organisation. And that, in itself, is almost impossible for an adversary to attack and defeat because it has no physical manifestation that can be truly destroyed.

Such an analysis might not assist military personnel in constructing a brilliant battle plan, but it should allow us to identify those things which provide the fundamental capability to win in combat and warfare. Having identified such fundamental capabilities a natural response would be to both protect and invest in them. A military force that really understood this would be visible in their investment in their people: ways of working, talent management, training, professionalised education, leadership development, mentoring and education. Unfortunately, there are not many militaries that subscribe to this view, preferring instead to invest more in technology and less in people, or in human skills and attributes. The educational and training budgets for military people are given a tiny percentage of investment in terms of time or resources from leaders today. It is no exaggeration to state that leaders care less for considerations of morale, engendering initiative and freedoms, investing in education and professional development of their people, that they do for accountancy training, data science projects, or efficiency measures.

It is this issue that Professor William Scott-Jackson addresses. As a leading scholar and researcher into human resources, his views of industrial and commercial organisations translate surprisingly well to military institutions.

In thinking about the 'Western way of war', there is a danger of technological 'presentism'; a belief that today's technology – of something promised within a few years – that could provide a decisive edge in future conflict. Commercial entities and innovators have been promising this to military forces for millennia, but hitherto senior commanders and leaders have maintained a healthy degree of cynicism to such promises. Today, however, it would appear that many people in the profession of arms have been simply accepting the hyperbole of the technology sector that they will be able to (or have already) provided the keys to capabilities that will be the decisive factor for future conflicts.

In wondering why this narrative about technology seems to have become ubiquitous across the officer cohort of Western militaries, one should consider what skills and knowledge is being imparted during periods of professional military education that overturns so much of their own experiences on operations and exercises, and leads officers with decades of experience to start talking and thinking in a way that is completely alien to everything they had seen in their careers thus far.

During many of the discussions in recording the podcast series, a myriad of guests talked about how the current offering in professional military education is not fit for task, despite the more or less constant reviews of it and tweaks to the syllabi that seem unending. Indeed, more than one guest has remarked that the UK, like many European militaries, seemed to be unable to construct and engender an environment that creates great military leaders despite the investment of time and resources put into various courses. One critique often raised was the idea that Western professional military education was designed almost entirely for the most promising officers – perhaps (being generous) 5 per cent of the total force. Little effort or resource is expended in the remaining 95 per cent. Isn't the lance jack as likely to be as decisive on the ground as the corps commander in his beautiful headquarters. Andrew Porteous, of EthicEyes, has done some interesting research about talent and strategic or distinctive advantage, and he put me in touch with Professor William Scott-Jackson, to provide some more context to this part of the discussion.

Named by Chartered Institute for Professional Development as one of the top five thinkers in human resources, Professor William Scott-Jackson is an academic and consultant researching and advising countries and organizations on maximizing their strategic capabilities. He's associate fellow at the Oxford Centre for Mutual and Employee-Owned Business at Oxford University and a visiting professor at Cass Business School [NOTE: It changed its name to Bayes Business School in 2021.] His recent book on *Transforming Happiness, Well-being and Engagement* has been extensively cited, and he's the author of the Chartered Management Institute's Guide 'Learning to Lead' and 'Managing Human Capital' and the CIPD guide to HR outsourcing. William's an academic advisor to the Chartered Management Institute's Human Capital Working Group, a member of the CMI's Academic Advisory Board, and the CIPD Professional Standards Faculty as well as the national advisory board on 'Valuing Your Talent'.

What Does the Western Way of War Mean to You?

William Scott-Jackson: I guess the key thing is about evolution. It has obviously evolved from a fairly predictable set piece – strategic engagement between two well-defined sets of people – to a much more unpredictable, virtual, less controllable, continuous engagement between almost every political entity at varying degrees of hostility. It has become much more messy and much less predictable, and much less, I suppose, controllable, by the forces involved.

Peter Roberts: I think that's a pretty accurate description of what we've seen militaries engaged in. There's the great spaghetti diagram that [General] Stan McChrystal came up with when he went into Afghanistan, that tried to illustrate and articulate the number of players and people who have an interest in the campaign, and how he (as the commander) saw the plan relating to them. It was impenetrable to try to understand. The problem with lots of military approaches to complex campaigns is that military personnel try to simplify it and find that single centre of gravity, that single spot on which they can turn the entire campaign. And they try to think that there's a way that if they undo this single factor, that they can undo that myriad of problems that are associated with. Now, that involves an incredible gift, if you like, but some serious intellectual capacity in the human involved. But that matters on the ground as well if you're a lance corporal out patrolling in Helmand: engagements can trigger around the responses that one individual makes, and in many ways there's a pattern. If we're seeing the evolution of war in this way, perhaps we haven't identified and grown our talent properly. I guess that's really why I want to come to you. The misidentification of your strategic advantage is pretty dire, correct?

WSJ: Yes, exactly and what kind of capabilities you need to maximize that advantage is also very rarely worked out, not just in a military, but certainly in most organisations. The link between the strategic advantage, how we compete, and the capabilities we actually need to do that are not made at all clear. Most people focus on what you might call fundamental capabilities. As we know in the military there's cyber capabilities, there's technical capabilities, there's numbers of ships, types of ships, all sorts of stuff. But those fundamental capabilities are the same as everybody else. You might be a year or two ahead or a year or two behind, but you're basically just playing catch-up. These fundamental capabilities can never provide an ongoing advantage or a competitive advantage to any organisation, including the military. What you need to look for is what's called strategic capabilities, and these are ones that provide a long-term advantage. If there's anything about the military that I would guess is true it's that you're always in a competitive situation. You're needed to beat somebody.

PR: That's a really interesting way of looking at platforms and equipment we have around us in Western militaries; as these fundamental capabilities. We need them to fight, no one's doubting that. We need bombs and bullets, and we need medics and we need ships and submarines and fighter aircraft and drones and all that stuff, but – as you say – that's never decisive. Even if you have an advantage in one area, or perhaps in many areas, it's for a fleeting period of time before you're playing catch-up with someone else. So these strategic capabilities aren't any of those things, right?

WSJ: Correct. What they [strategic capabilities] are, in a formal sense, are things with something that's valuable towards your strategic advantage. So the fundamental things we've described are all valuable, but these strategic ones have to be valuable as well. The reason I mention that, in particular, is because they also have got to be very rare. So, not everybody's got them. But they've also got to be what's called inimitable, which means that it's not easy to copy or build them, and this is where it gets quite interesting and tricky. They have got to be rare, hard to copy and valuable. So, for example, a military cyber capability isn't particularly rare. It's fairly easy to copy, although it is valuable. It fails on those criteria. We have to look for something else and it turns out, of course, that technical skills of any description, or technical stuff – equipment or anything else – generally speaking fails on the criteria, because it's always copiable, and this is why you're always playing catch-up. It turns out that strategic capabilities are usually human factors because they're much more difficult to define, they're much more difficult to create and they're quite often hard to copy. In something like the military, they are quite often based in the previous culture or the history of the organisation, which makes it even more difficult to copy. You could say, for example – I don't even know if this is true – but you could say the particular culture of the British Army is that people use their initiative. Well, where did that come from? Was it deliberately built? Was it a sort of set of training courses? Or is it how it has always been and people have come in and been indoctrinated into that way? These human factors tend to be much more difficult to copy and bizarrely they're very rarely focused on. So going back to companies, for example, we did a lot of work for a bank. My main advantage as an academic is I do a lot of consulting in the real world, so we did a lot of work for a bank, and we found out that their key strategic capability was the friendliness of their staff. When you think about it, if you ask the bank, if your people were 15 per cent more friendly than everybody else's would that give you a profit? They would absolutely say yes and then you go and look at their HR processes for recruitment, training, development or anything else, and it doesn't mention the word in any way. People are brought in because they're efficient, because they're whatever, they can add up, they can do whatever they need to do. Then you look at the processes and the recent processes in this particular bank include things like targeting people on the number of customers they see per day, which completely argues against those people being friendly. So not only did they not know what their strategic capability was, but all their processes, their recruitment, their training, their actual way they worked mitigated against that very capability. They were busily engaged in destroying it.

PR: It seems like this requires an understanding of the culture and what you want out of your workforce, right?

WSJ: Yes. One of the key issues is that although even in the military people will talk about morale, enthusiasm, initiative, whatever, these terms are seen as kind of floppy. Yes, they are great but what actually is it? Friendliness, if we were talking about this bank again: what we ended up doing was doing a lot of psychology work to define very scientifically what friendliness was and what it meant in their particular circumstance. It didn't mean chatting with somebody for four hours; it meant being open and whatever. You can define these terms a lot more tightly than they're normally defined, and once you've done that you can look for things like how do you assess it properly: a scientific way of assessing this particular factor. Is there a scientific way of building it [the skills and behaviours you need] in people? Is there a scientific way of recruiting for it? Once you've done that, then you've got a human resource policy which is actually dedicated to building that strategic capability?

PR: That I find absolutely fascinating because – as you say – if you look at morale, and the Armed Forces Continuous Attitude Survey recognises there's been this tailing off of morale in the past decade, that military members and their families both feel they don't get the reward, and you get senior officers and politicians and civil servants continually engaging and claiming to be on top of this problem, but exactly as you say I'm not sure they understand the science behind what morale means. For them, it's a real, sort of, fluffy, furry thing which we don't really understand in a scientific way. Maybe we give people some more sport and then maybe morale will improve. But I suspect it's more complex than that.

WSJ: Absolutely. There's a similar kind of concept in business, which is causal engagement. This is the degree to which people put in discretionary effort to really do a job even though they've finished working for the day, or whatever. It's fairly similar, in a way, to that. Most organisations say engagement is really important and most organisations have got a plan to do something about engagement, and all these plans include things the company does top down. Increasing wages, dealing with pensions, having social events, all sorts of stuff. When we did our investigation [at the bank] and we found out that successful engagement actually is at least 40 per cent due to the person themself. It's their propensity for engagement that is one of the most important factors, and yet nobody recruits for engagement. Nobody develops engagement in the person as a capability and the factors they are playing with are not the ones that actually affect it. The same with morale. I haven't studied morale but if I did, I would definitely find that most of the levers that the military is using to bolster are probably not the levers that actually affect it.

PR: Just thinking about what you're saying, could you actually understand and test for it these key facets of a person in the recruitment process rather than testing for a soldier's propensity for violence? Would it be more helpful to test for a propensity for restraint, for example?

WSJ: Exactly. ·

PR: … Or rational thinking and decision making, and you could test for that at an early stage [of the recruitment process] and thereby understand how you would best exploit individuals and where they might excel within a military?

WSJ: Exactly, and because especially with the military, you have to recruit a lot of people, so you'd have to take a fairly wide range even if it was on those factors that you just discussed. For example, you couldn't take only one per cent of the population because they happen to have those things. But you also need a way of developing it in other people, and those are the two main things. That's basically what human resources is supposed to do. It's supposed to build, recruit and develop strategic capabilities. But it is also true – sad, but true – that this is not happening often.

PR: It's surprising that the military, in many ways, is very good at training leaders. Less good at training managers, and I get the impression that the military is pretty poor at executing strategic-level HR. When you start reading about this stuff and some of your research, the reality is that the military practices just aren't that good. The way that militaries appoint people and just put them where they need to be, rather than examining the two-way relationship between people and their skills. These are common problem sets that you get in lots and lots of companies, but isn't it true that many companies are now acknowledging that this is the wrong way to do things?

WSJ: Yes, exactly. It's interesting that you mention leaders because, again, we've found that with engagement the next most important factor is the immediate leaders. But it's not the top person, it's not a brigadier or somebody. It's my immediate squad leader, the person I see every day. That's the one who really influences me, and that's the one who can destroy or improve my morale. Therefore, doing things like pay rises or increased time off or something, or better housing, or big events or anything else is not addressing this issue. The key issue is that 'my' leader needs to motivate me and we need to train our leaders how to do it. So for leadership training, if you thought morale is the key thing, which we haven't agreed yet but it might be, then you need to train the [junior] leaders, because they're the ones that affect it. But not training them to make good decisions isn't the key point. The key point is how do they motivate and keep the morale up of their little squad. It's got to be the first-line leaders. It's the corporal; it's not somebody higher up.

PR: I think this is the critical point that we're talking about, because in the British military, and I think in most European militaries it's the same, we tend to associate leadership with officers.

WSJ: Yes, it is the same in business.

PR: It strikes me from what you're saying that this is completely wrong. Actually leadership is performed by the lance jack, by the corporal, by the sergeant, by the chief petty officer or the petty officer. This is the level which makes the greater impact on your ability to wage war – how they lead, motivate and run their teams. Would you subscribe to that view?

WSJ: Indeed, yes. The good news is, if we just talk about leadership, which is one of the factors – it's a thing, it's not everything about it – but if we just talk about leadership, again it turns out that most people that get promoted to a leadership at an early stage, like a corporal, they're trained in all sorts of stuff and they get a lot of training and it's quite complicated and they remember a quarter of it, but the bit they really need is very simple. It's that how do you tell somebody off when last week they were your mate? There's an actual process you can use, which works every time, whether you're the most charismatic person in the world or whether you're not.

It's a very straightforward process, which you can teach a new leader in about half an hour and give him a guide. How do you change everybody's direction, really quickly? How do you have a meeting that isn't really boring and sends everybody to sleep? There are 10 things that we identified in commerce – I'm not saying they're the same in the military but they certainly in commerce – and if every leader had one day's training and was given a little guide, preferably an app, that said this is how you do these 10 things, then they all would be extremely effective leaders, irrespective of their personalities, their accretion, their charisma, their authenticity or anything else that's complicated. One day, and all of your leaders would be effective.

PR: That's phenomenal! Because I have a huge amount of respect for the junior ranks: the corporals, the petty officers, the leading hands. These are phenomenal people who do jobs under extreme amounts of pressures who don't really get a lot of recognition, who I also believe could run meetings with a lot more humour than their officer counterparts, and some of them do get some long periods of training. For the Royal Navy, the leading hands leadership course is a decent period of time – and quite tough. The petty officers leadership course, again, is significant. For the Army it's the same, and for the RAF I'm sure it's likewise. Militaries invest some time in these people on promotion but I'm not sure that the syllabus is geared towards how we skill them correctly. Instead, the majority of military training in leadership and management is based on the theory of action-centred leadership. The sort of dictatorial measures. I'm not sure how useful they are compared to the skills that you're talking about, because we [militaries] don't think about leadership or management in that way. The leadership and management that we think about in the forces is centred on how officers lead large groups of people. It's not centred on direct engagement with small groups.

WSJ: That's very interesting, because that's exactly mirrored in civilian life. If you're a new leader then it might be that last week you were one of the guys on an oil rig, just doing your stuff, and all of a sudden somebody said to you, right, now you're the supervisor. And you've got 10 people, and you're supposed to be the supervisor. If you do get training, it will be training based on things like how Bill Gates ran Microsoft or what Steve Jobs did to inspire the company and provide a vision. It could be a six-week course in how to be a charismatic leader or how to be an authentic leader. And you've got no idea what's going on. It's all very interesting and stuff, but you won't be able to remember it, you certainly won't use it, and you're never going to be Bill Gates anyway, because you're not like Bill Gates or you're not like Steve Jobs, and actually they were not necessarily good leaders either of individual people. What you want is the basics, which is how do you do this job more effectively in a very simple way. That's why we say boil it down, and basically we believe – and as I say we haven't looked at the military, so I'm not sure if this is true in that case – but certainly in civilian life, for a supervisor they need to do 10 things really well. And those 10 things could be written on 10 bits of paper. And they make a hell of a difference.

PR: That's fascinating. I can remember some senior officers who need to do 10 things a bit better in terms of leadership as well. I don't think it's exclusive. But one of the interesting facets of this is the culture of the organisation, how you imbue that. For example, regiments in the British Army, Royal Navy, individual ships, platforms, units, squadrons – they have a culture of their own. They have a sort of ethos that sits deep within it. Most of them are really good. They're really productive. It's competitive with their peers as well as the adversary. In this manner, they generally drive themselves to better performance. It's what they do. It's why they almost exist as units. That culture is not that different between each part of the armed forces or between different platforms, but they are slightly differentiated and they take quite a long time to acclimatize to. When you're joining a ship and you think you know what it is because all ships are a bit the same, you join and it might have a few peculiarities, but you think you slip in. Yet over a period of time there's an ethos, a culture to it that is slightly different, that shapes you towards it as well as it towards you, if you see what I mean. And that feels a really important part of what you do. Now, part of your ability to adapt to that culture is the fact that you've had so much experience beforehand across a whole variety of military tasks that takes you there and makes you work. And I guess that's the same in industry. You can swap from one company to another company. The ethos might be different but if the product is roughly the same then after a while you'll adopt that ethos and you'll get there. Is that right?

WSJ: Actually, I would have said the military is far more homogeneous than companies in terms of culture. Companies do have very distinct cultures and people don't fit; they can be really having a bad time if they stay in a company that they don't fit in for any great length of time. I think the military culture is much more homogenous, probably, between sort of units and that culture, one way or the other, is going to be a source of this advantage, whatever that advantage is. Let's take a semi-ridiculous example, let's suppose we thought that the biggest thing for the military in terms of capability was the adaptability of its people, let's say. There's usually five or so, four or five things, that are critical or endow advantage. Let's say adaptability is one of them. If we look at the culture, and we say how does that culture reinforce adaptability? How do we make it so it works better, because everybody's doing the same kinds of things, and then you can actually recruit people that are going to fit more in that culture and provide that capability because the two things are bound together? If you found somewhere where the culture was actually against the capabilities, so supposing one of the units was extremely disciplined, everyone was very disciplined, assuming that isn't a strategic advantage for them because of their particular role or something, then you might have to try to do something about the culture. That would come down to senior officers and that would take some time. But generally speaking what you look for is something that actually fits with the culture, because that's where it's going to have come from, and it also makes it very difficult to copy, because you have to build the whole culture in order to have that particular kind of adaptability or that particular kind of morale.

PR: That's where I was – one of my first takes that I made on this, when you're talking about valuable, rare and inimitable – the inability to copy it – lots of the talk in modern employment models for the armed forces talks increasingly about the ability for civilians to jump into management positions from industry with no experience in the military whatsoever to jump in and suddenly take over in those positions and then jump out again. I'm wondering how you think that would play against this idea of culture being inimitable, that is grown within you. How is that likely to work if people and this culture, this sort of system of deep knowledge and understanding, is the decisive edge?

WSJ: If you're bringing somebody in at a management position because they've got a technical skill which you need and you can't get cheaper so you have to make them a something, but they haven't actually got any management responsibility, which I would guess probably happens sometimes, if it's that, there's no problem, really, because they're just doing it like a contractor. They're like a contractor. They're just doing their job, no problem, unless 60 per cent of the unit was contractors, in which case the culture would definitely change. But if they're actually leading somebody, then they are going to be starting to affect the culture with the way they do things, so you'd either have to include that as part of the recruitment package – this person – however we define the culture and however we define the capabilities, and it would be in detail that would be scientific, if you like. Then you'd have to recruit against those criteria as much as you would against the technical criteria. Just because you happen to be the best cyber security person in the world, if you're actually going to lead a unit, we also need you to be the best adaptable, or the best high morale maker, or whatever it might be. That's the only way to do it. If you bring people in that don't fit, they will start to change the culture, certainly of those beneath them, and you'll lose an advantage. It's quite incredible, actually, the number of times – I mean, the British military will have, right now, some strategic capabilities that are absolutely unique, really hard to copy, really valuable, and are being thrown away by actions that are being taken by not realising that they're there.

PR: Everyone in the military will have stories about this kind of thing happening to them. There are chefs who are working in cyber who are brilliant but can't be advanced because they need to do a professional cooking course or something ridiculous, so we can't reward them and keep them in the area – there are loads of examples like that! Everybody in the military has them. But it's such a familiar occurrence that we think little about it. It is striking to hear it professionalised and scientifically analysed as you do.

Culturally, there's a sort of British culture, which underlies the British workforce, which underpins the British Armed Forces: it makes the military reflective of our society. Is the same model transferable across other states and cultures, so the French will have one and the Germans will have one and the Russians will have one. Will they all be slightly different?

WSJ: Absolutely and the good news for the British is that they'll be equally bad at recognising what it is and they'll all be busily destroying it as we speak. Whatever distinguishes the Russian military machine will be being affected by the decisions being taken and it will be affected adversely in terms of strategic capabilities. I could almost guarantee it.

PR: That's my next point. If we know these things about the adversaries, we can accelerate the decline of their decisive or strategic advantage by understanding it and then playing to it, right?

WSJ: Exactly and so, for example, if you know that every force in the world valued morale, and they were all doing the same things then it can't be a strategic advantage, so we can do something else. You know we need to do another thing, a different thing. Because this difference thing is absolutely crucial. It can't be the same as everybody else.

PR: So not only can you increase your own decisive advantage but you can negate your adversary's strategic advantage if you do the background, the research, the understanding, but it also comes out of this idea that actually you can't do this to a fundamental capability, you can only do it to one of your strategic capabilities, and usually most of those are human, right?

WSJ: Absolutely. That's a very good summary.

Reflections

There are number of key deductions from this discussion. First is that reliance on a centre of gravity analysis alone is likely to be flawed in understanding the fundamental capabilities of your own military: those things that one should invest in and nurture in order to exploit. Second is that in developing these skills, real investment is needed – underpinned by the appropriate research – to maximise the lethality and prowess or those fundamental elements of your military force. Third is an understanding that the most effective investment to gain increased fighting capability is rarely at the top of the organisation, nor is it at great distance from the fighting human. Instead, investment (of time, energy and cash) will deliver greatest efficacy if provided in direct proportion to distance from the working end of the military. Ensuring the welfare, morale, and fighting spirit of front-line soldiers, sailors, aviators and their combat support, maintenance, and service support comrades does not deliver beautiful metrics that a senior officer can measure. Yet it is this investment in the people who will do the fighting that will yield the greatest results when war arrives. Those people are valuable, rare, and immutable: one should think very hard before considering a deliberate action that will change their culture, ethos, and values.

4

Nuclear Weapons

Modern military forces, in their make-up, culture, ethos and values, broadly seem to represent the societies they are drawn from. Military organisations can shape some of these with their own institutional culture and experiences through training, education and exposure, but the days of states employing non-representative military forces (for example in the use of mercenary forces by Ancient Greek city states, or of the Medici era pre-militia forces of Italy), seems to have passed – at least for the moment. It therefore follows that as societies evolve in their ethics and values, so too will the military force. Behaviours, expectations and presumptions also transfer into the military force. The nature of such organisations, in that their ability to make dramatic, fundamental and irrational shifts in outlooks are rare outside of transformative experiences as with a major military defeat. Yet, over time, they do adapt, morph, change and evolve.

However, not all military forces, nor their parent societies evolve in the same way or at the same rate. There is no homogenous path down which all militaries evolve; the same is true for the societies they represent. This can be problematic when change occurs at different rates and in different ways for forces that co-operate most closely. It is an even more significant problem when adversary states take entirely separate pathways or cease to evolve in certain areas or ways. This is accentuated and amplified when states make assumptions about the culture, beliefs and behaviours of states (a presumption underwritten in most Western National Security policies). The notion of China's 'peaceful rise' is a clear example of this: an idea of the 2000s that the Communist Party of China would oversee a smooth integration of the world's fastest growing economy (and population), into an existing world order governed by a set of rules and behaviours that it (the CCP) had no hand in shaping.

Military assumptions can be no less profound in their flaws. Since the 1980s the West has evolved a view on nuclear weapons such that by 2015 their use seemed incomprehensible. States and their populations in the West regarded the devastation reeked by such weapons as so irrational, illogical and nonsensical that use could not be imagined. Indeed, the consequences of the use of any nuclear weapon was deemed to be so devastating on the wider world population that some militaries started to distance themselves from those capabilities. This shift in thinking about nuclear capabilities took place as the last generation familiar with Cold War doctrines of nuclear use left active service in Western armed forces.

Throughout the Cold War, both NATO and the Warsaw Pact possessed and exercised doctrines designed around the use of battlefield tactical nuclear weapons. Loading and flying with such weapons was practised on a fairly regular basis. Deploying tactical nuclear weapons on ships, aircraft and in the field was, if not commonplace, then certainly something familiar to service personnel, their commanders, and political leaders who were required to authorise such things. It would be hard to imagine a senior officer in the West who was not intimately familiar with circumstances in which they would be used, when deployment was authorised, and the exact timing of use was delegated to subordinate commanders.

Post-Cold War there was a distinct attitude shift in Western states away from weapons of mass effect and towards precision. Tactical nuclear weapons were removed from war stocks and capability lists from the majority of NATO members. This was not the case in Russia, where such capabilities continued to be fielded and exercised with, specifically on their Eastern and Southern borders with China. In purely martial terms, tactical nuclear weapons were the only cost-effective weapons able to overcome the potential mass that the Chinese might bring to bear in a Sino-Russian conflict. Yet this factor was largely ignored in the West, and the presumption emerged for Western military and political leaders that whilst nuclear capabilities might remain a core element of deterrence against similarly armed powers, their use was considered an extremely remote possibility. As such, even European states that continued to possess nuclear weapons ceased to conduct exercises or to qualify people in nuclear planning.

One might be surprised at this, particularly given the development of Iran's nuclear programme over this period, the mooted possession of nuclear capabilities by Israel, and the much-publicised evolution of a nuclear capability in North Korea by successive leaders (along with the means to deliver them). The hubris in European capitals was staggering regarding each of these potential threats. There were few discussions regarding nuclear weapons across Whitehall in London (the decision-making centre of the UK), for example, and those that did take place were poorly attended and often rescheduled at short notice. There is little evidence that things were different in Paris, or in states who had previously considered the use and transportation of nuclear weapons provided by other Alliance members (Germany, Italy or the Netherlands).

All of this can only be chalked up to a lack of imagination from political and military leaders who all considered nuclear matters to be beyond imagination. Indeed, between 2010 and 2020, it is challenging to identify a single serious event on potential nuclear weapons use (war-gaming, exercising, round table discussions, conferences, or seminars) that could attract a senior audience and have an informed discussion.

The following discussion took place on 09 December 2021. To put this in context, over the previous year two Scandinavian states had requested nuclear guarantees from the US, and various discussions in the Baltic states and Poland had been airing the issue of Russia's use of nuclear brinkmanship. In addition, after 2019, there was an increasing interest from the academic community in the US regarding China's nuclear doctrine.[1]

Dr Matthew Harries was the foremost independent nuclear researcher in Europe, and a Senior Research Fellow at RUSI based in Berlin. Matt had worked on UK Parliamentary committees, as well as the managing editor of a peer-reviewed journal, Survival, and as an academic at the John Hopkins University, and at King's College London.

What Does the Western Way of War Mean to You?

Matthew Harries: I come at this question differently compared to the distinguished warfighters and thinkers about war that you usually have on the podcast; I come at this question sort of upside-down because I spend

---

[1] See, for example, Gerald C Brown, 'Understanding the Risks and Realities of China's Nuclear Forces', in Arms Control Association commentary, June 2021, https://www.armscontrol.org/act/2021-06/features/understanding-risks-realities-chinas-nuclear-forces, or Sari Arha Havrén, 'China's No First Use Policy: Change or False Alarm', in RUSI commentary 13 October 2023, https://www.rusi.org/explore-our-research/publications/commentary/chinas-no-first-use-nuclear-weapons-policy-change-or-false-alarm.

my time thinking and writing about nuclear deterrents and nuclear arms control. So, for me, the Western way of warfare is based on the idea that unrestrained war between nuclear-armed powers is unthinkable. That means that any kind of war between nuclear-armed powers is extremely dangerous, and because the nuclear powers are the major states, by and large in today's international system, that means we have a world run on the idea that nuclear war would be so awful that we can't set a foot down that path. There's a phrase that Presidents Reagan and Gorbachev put down on paper first in 1985 and which Presidents Biden and Putin reaffirmed [in 2019], which is that a nuclear war cannot be won and must never be fought. I think most ordinary people would agree with that sentiment, but there's an inherent problem with that, which is that if you're totally unwilling to use nuclear weapons, if you're totally unwilling to fight nuclear war, or at least a war that goes nuclear, then there's no such thing as deterrence. So, we've always been caught in this trap which is, how do you make credible what seems like an incredible threat, to use nuclear weapons?

Peter Roberts: I think that's a really useful, and as you say, different perspective because most people come at it from the idea that the end result of a nuclear war would be accidental. A conflict would start small and escalation – intended or otherwise – could end up in a conflagration that resulted in the use of nukes: A sort of Thucydides Trap [that war is bound to happen between the current great power and the rising power] and coming at it in your way puts a very different spin on it. This paradox that you present between what you want to believe, that it can't happen, and the reality that you must be willing to do it, sits right at the heart of every problem we see people struggling with whenever they engage with this debate. The contrast between debate in the US and Russia, for example, against the discussions in Europe is stark.

In some parts of Europe, and it's not just the UK, discussions about the possession – let alone the use of – nuclear weapons are rarely conducted in public. In Germany, for example, the debate over what a US withdrawal of the nuclear guarantee might mean makes headlines but discussion is rarely more than skin deep. Within such discussions there is a very liberal ideology that makes it hard, as societies have evolved, to engage in a discourse about nuclear weapons and nuclear war, that seems so at odds with how we hope the world works.

Neither adversaries nor many allies think about nuclear issues in the same idealistic way; other states have more serious conversations about what it would take to deliver credible nuclear deterrence in the Nuclear Age. Americans and the Russians actively exercise and pursue nuclear training across their militaries; politicians of these states discuss tripwires, talk about requirements of how and when they would fire [nuclear weapons]: engage meaningfully in conversations about the ethics and the boundaries to which they go to. These activities happen in the US, and in Russia, and perhaps in China (who knows). But they are largely absent in Europe. I'm not quite sure about France, but in the UK, it feels like we don't have those conversations.

There's only one nuclear planner in the UK who trained in the US. That's slightly incredible for me, as someone who examines military theory, that you possess these things, but really the military don't want to have anything to do with them, and the politicians really don't want to have anything to do with them. We don't have an active debate like in Europe about nuclear weapons, really. Do you think I'm only seeing one side of the argument? Do you think there is something there, but in more specialist areas?

MH: I think things have changed for sure. To put it in simplistic terms, there's always an argument about nuclear weapons and arms control, where one side of that argument thinks that if you pursue arms control, you'll reduce the threat from nuclear weapons and you'll improve relations between the big powers. The other side of the argument says that you'll get arms control when relations between the big powers get better. Are nuclear weapons a cause of problems, or are they a symptom of the problems? I think what's noticeable, and where I disagree with you a bit, is that actually, since 2014, since Crimea, NATO-Russia relations have gone so rapidly downhill, and so seriously downhill, that actually nuclear deterrents and nuclear weapons are higher up the agenda, much higher up the agenda in fact, than they used to be, and I think that applies to public debates as well. To a certain extent, publics don't care about security policy at all, right? So, any worries that we might have that this is not high up the public agenda equally apply to what the public thinks about military modernisation, what the public thinks about a lot of things that we're interested in, and we want everyone else to be interested in. You mentioned the UK. I mean, the UK's Integrated Review (in March 2020), marked at least a kind of psychological tipping point.

It got a lot of headlines because the UK raised the cap on the number of warheads the UK says it reserves the right to build, which is important in itself, but actually in terms of message sending, in psychological terms, what the UK really did was to say, 'We are in, and we're recognising we are in a world where nuclear weapons and nuclear deterrents are more important.' I think that realisation is sinking in. I think in Europe, I mean first obviously, there is no such thing as a single European perception on these issues, it very much depends on where you sit, and the feeling about these issues in the Baltic states, for example, is very different from parts of Western Europe. But even in Germany, where traditionally there has been an aversion to talking in security policy language at all, and where the majority of the public really do not like nuclear weapons, actually you're starting to see, I think, a gradual shift in perceptions. Not necessarily to become more enthusiastic about nuclear deterrence, but to realise the kind of changed environment that we're living in. So, we're talking on November 25th (2020). Yesterday, the new German coalition, the so-called traffic light coalition of the Social Democratic Party, the Green Party, and the Free Democratic Party, released its coalition agreement, which people like me had been hotly anticipating to see what it would say about nuclear weapons because both the SPD and the Greens are historically fairly anti-nuclear parties and have a large part of their party base which would like to see nuclear disarmament accelerated, which would like to see US nuclear weapons withdrawn from Germany, and Germany to stop participating in NATO nuclear sharing arrangements, and would like to see Germany support the new nuclear ban treaty, the Treaty on the Prohibition of Nuclear Weapons.

Actually, the coalition agreement says that Germany will go to the first meeting of the parties to the ban treaty but only as an observer, and it says, in slightly cryptic language, but basically, it says that Germany will procure a replacement for the aircraft that currently carry nuclear weapons, the Tornados, and that the replacement will be certified for nuclear use. So, in practical terms, despite the instincts of some sections of its component parties, it doesn't look like the new German government will change course very much on nuclear weapons. Nuclear disarmament and arms control will still be an important goal, and Germany will still want to contribute to multilateral processes to advance that goal, but actually, I think there's a realisation that now is not the time to split NATO on these issues. So, there's a sort of reluctant acceptance, I think, that this is the world we're living in. Part of that is if you look at the rest of what, in particular the Green Party for example, wants to do on foreign policy and security policy, they're actually quite keen on quite a significant toughening up of German attitudes towards Russia, and towards China. In a way, the traditional emphasis on nuclear disarmament would have had the potential to come at odds with that. If you decide to get rid of US nuclear weapons on German soil and get out of NATO nuclear sharing, you're precipitating a big fight. Whether or not you think that US B61 gravity bombs on Tornados or a successor are actually going to be any use in any plausible scenario is, sort of, beside the point. The point is that the fight that you precipitate within the Alliance when you do that would be a big deal and would get in the way of toughening up that policy on Russia. So, I think in essence you're right to say that the theology of nuclear deterrence possibly isn't at the top of everyone's minds, and certainly not the general public's minds, but actually, I think there is a growing realisation that we are now likely to be living with these weapons for longer than perhaps we'd hoped, and that's a realisation that leads to a whole bunch of difficult decisions.[2]

PR: If we've covered off Europe in those ways, and in saying that actually the debate is becoming more sophisticated, it's higher up on the agenda as we move east across the world, out of Europe, towards Israel and maybe Iran and Pakistan and India, is the debate about nuclear weapons usage there very different from what we'd see in Europe or in the United States? Is it evolving in the same way? Is it as much alive and as interesting, and as engaged within both political and military terms as we see in Europe?

MH: Well, I think the first thing to say is there are obviously different perceptions of nuclear weapons in different parts of the world, it's hard to generalise. But I think the one thing you can say is that unfortunately, there are no nuclear relationships in the world today that are getting more constructive rather than less constructive. There are no relationships between nuclear-armed states, where the nuclear weapons are becoming less relevant, rather than more relevant. In a way, going back to this question of whether nuclear weapons are symptom or cause, in a way the West is catching up to the fact that when you are genuinely worried about the prospect of major war in a way that I think the West hasn't been for

---

[2] As a result, the political debate over whether the UK would perhaps provide a nuclear guarantee to NATO members, or perhaps some European neighbours has been mooted across the continent. See, for example, Patrick Wintour, "UK could contribute to nuclear shield if Trump wins, suggests German minister", in *The Guardian*, 15 February 2024. https://www.theguardian.com/world/2024/feb/15/uk-europe-nuclear-shield-donald-trump-germany-nato-deterrent.

quite a long time, then you have to think seriously about nuclear weapons again. So, in South Asia, for example, nuclear competition or changes in the nuclear balance between India and Pakistan have been very important for two decades and have not been in the background. They've been in the foreground, and it's been a live debate, and the deterrence strategies of those countries have been constantly shifting, and constantly being updated. For Israel, obviously existential considerations, and nuclear weapons are about existential considerations, have never gone away. For North Korea, you can say much the same things, that the ability not to have to think about nuclear deterrence as a preventer of major war is something that, in East Asia, hasn't been a luxury that they've been able to afford. So in a way, I think we're catching up.

PR: If everyone's catching up, then the recent images of the massive increase in preparations that China has been making as a nuclear power are driving again a huge number of changes in the US and elsewhere that I want to get to in a minute. But on China's nuclear programme, we have just seen a really significant rise, or a public revelation that the number of land-based silos has started to increase and exponentially change the nuclear balance of power. What do you make of China's understanding of nuclear weapons within that doctrine? If previously we've really understood how Russia and the United States thought about it, do we understand how China thinks about nuclear weapons? Does it think about it the same way or is there a different dynamic in there?

MH: Before we start using words like massive and exponential, I think there's a tricky aspect to this which is that things are coming into the public domain gradually and I think our understanding in public of what is actually happening in China is lagging behind what is actually happening, and what presumably the US government has discovered, or perhaps the UK government knows. This is a tricky area because I think at the minute we're working on incomplete information. Even at the minute, the gloomiest scenarios for the US defence department's China Military Power Report talked about China getting to at least 1,000 warheads by the end of the decade, and that is on top of what it previously thought.[3] They were

---

[3] According to the US DoD 2024 report, at the rate anticipated growth, China will have a stockpile of 1,500 warheads by 2035. For 2024 estimates of the Chinese nuclear stockpile see, Amrita Jash, "By the numbers: China's nuclear inventory continues to grow", in Lowy Institute, 27 February 2024. https://www.lowyinstitute.org/the-interpreter/numbers-china-s-nuclear-inventory-continues-grow.

previously talking about at least doubling China's stockpile, and that was doubling a stockpile in the low-200s. So yes, the forecasts for where China's going have increased really significantly. On the other hand, and yes the revelations of China building what looks like at least three new ICBM silo fields with hundreds of potential silos, is a big change, and yes the reports of a hypersonic glide vehicle test that was put into orbit first, and which, according to the latest FT story, then fired a missile from the glide vehicle while it was gliding. Yes, this is all very important, and very significant, but we just don't know, I think, where it's heading yet.

There are some people who theorise that China might not be on a full-blooded crash programme to increase its numbers as quickly as possible, that it might keep some of these silos empty, and you always have to keep a critical eye on who's leaking what and why. I personally don't think that, for example, the reporting on China's hypersonic glide vehicle test has yet properly explained what the United States thinks happened and why it's significant. That's not something I blame the reporters for, I just don't think we have the full story yet so I'm remaining cautious. All of that said, however, and while it's important to be careful with your adjectives, especially if those adjectives have to do with numbers, I think what we are seeing is a shift away from what we thought was the way that China thought about nuclear weapons in the past, to something new. It's not clear what the new thing is, but what you definitely have, even if we don't know what the trajectory of the numbers is going to be yet, is a very serious and committed diversification of China's nuclear arsenal. So, more sophisticated missiles, more of them, ballistic missiles, submarines going to sea, the air force getting a nuclear role again.

So, you've got diversification, and you also seem to have innovation happening at the same time, what that forces us to confront basically, in my opinion, is the realisation that even if it was true in the past that China was willing to stay out of any kind of nuclear competition with the United States, and even if it was true in the past what China cared about was assuring retaliation to nuclear attack and that China would not use nuclear weapons first, something is changing in the way that China thinks about nuclear weapons now. It looks like that even if China's not going to pursue numerical parity with the United States, it might be less willing to be clearly inferior across the nuclear spectrum. It looks like China might be less willing to exist in a situation where it's clearly outmatched in various domains of nuclear competition. The reason I'm struggling with my words is because it doesn't look plausible that China can get to real numerical parity any time soon. I don't think there's the evidence yet that China's trying to outmatch the United States, and it's also always been true, I think, that the United States is in a relationship of vulnerability towards China. So, China's ability to respond with a strategic attack on the United States has always been there and in some way some of these changes don't necessarily affect that, that the fundamental of deterrence was already there, and might not be affected. But I think the question is, where's China going with this, and what does it want to change? What it might want to change is the options that it feels it has in the nuclear domain now that the prospect of conventional war between the US and China in the region seems more likely. The best way I've heard this summarised recently is to say that what it looks like China is doing is to take the option of US nuclear coercion during a conventional war in the region off the table. That's what China wants to do, but we don't know.

PR: I think that's one of the interesting parts, but as you say, the discussion about this in the US, and what's in the public domain, are very different things, right? So, what the US will know in a classified way, what it's making assumptions about in a classified way, which is not open to everyone as part of the discussion are clearly shifting some things. So, there has been a requirement for the US to embark on a mass recapitalisation of its nuclear triad, which it has delayed and prevaricated over, it's like a familiar theme in any defence decision, right, is prevaricate, delay. But there is this massive recapitalisation that's got to happen, and the arrival of China as a serious nuclear player has obviously changed some of the discourse in DC, and the prioritisation about what that looks and feels like in terms of their recapitalisation, and that programme is now behind where perhaps the US would want it to be, but it is still accelerating, isn't it?

MH: The context is that the Biden administration, at the time of speaking, has been aiming to have a Nuclear Posture Review done by the end of January 2022. President Biden is somebody who has, for decades, taken a personal interest in nuclear issues, and it's a subject where the president has opinions. It's an important review, not just because, as you say, there are some very important procurement and modernisation decisions coming, but also because it follows the Trump administration and I think the world is waiting to see in nuclear issues, just as in anything else, how much of the Trump administration was an aberration and how far back to the status quo ante the United States is going to go. The huge argument at the minute in terms of US nuclear posture has been over declaratory policy, what you say your weapons are for, and the buzzword in the US context is the idea of a so-called sole purpose declaration, that the US should declare that the sole purpose of its nuclear weapons is to deter adversary nuclear use against the US or its allies and partners. This is something that President Biden has personally supported in the past and supported in the campaign. The question is, is it going to make it into the Nuclear Posture Review? It looks much less likely that that's going to happen than it did before Biden became President, and I think the reason is, just to go back to what we were talking about before, that the role of nuclear weapons in regional security, and in the minds of the governments of US Allies, is increasing at the moment rather than decreasing.

US-China competition is becoming more nuclearised. The idea of nuclear deterrence is more central in thinking about the potential for conflict between the US and China now than it has been at any time, and the same thing is broadly true in the US-Russia, the NATO-Russia relationship. Basically, it looks like US regional Allies are not keen on the idea of trying to roll back what nuclear deterrence is for in their region. So, the worry is that if you say the sole purpose of nuclear weapons is to deter a nuclear attack then what you end up with is Russia, or potentially China, exploiting everything up to that line and thinking that there is no role for nuclear deterrence against anything non-nuclear. That's the concern. There's been a lot of I think private, but now also public, debate and signals of disquiet from US Allies about what that kind of declaration would mean.

PR: There we were, right at the start of this podcast, by saying we weren't going to touch on the deeply theological questions, and we end up at the end of time, the big questions are the heart of this, and actually, you can't get away with them.

Reflections

At the heart of Western ideas of war lie the assumptions about the ability of nuclear weapons to act as a credible deterrent for major war and conflict. Providing deterrence is the subject of a huge amount of scholarly literature that often can do little more than admire the problems, but nonetheless, nuclear weapons do form key parts of the military and foreign policy arsenals of the selected states who own them, either overtly or for some states that don't make it as public. We could argue about the ideas of nuclear deterrence as a concept, usually from our own views, shaped and informed by the state in which we operate, and usually markedly different from the rationality and logic that operates in other states. We might also debate whether the cost trade-offs between nuclear and conventional forces have the value that some believe. We might disagree about the viability of decision-making, political decision-making, the civilian command and control of nuclear weapons, their release authorisations, and the politicians' training for that role, about whether they exercise enough or at all.

We could enter into a passionate discussion about the independence [of nuclear arsenals], and what that means for nuclear weapons, and the sophistication of individual platforms. We should not ignore, either, those who want to accelerate the nuclear-weapon-free world with some unilateral decisions by states that own these weapons of mass destruction. And while discussion on each and all of these is valid, useful even, to inform and educate ourselves in the ways of war and our preconception of how others, enemies and allies alike, how they all view decision-making about these. There is some excellent research and analysis that has been published in specialist journals. But the aim of this conversation was to put aside some of those considerations and accept the fact that nuclear weapons will be a fact of life for some years to come, that their cost is an accepted element that secures our ideas of what it takes to retain what is commonly perceived to be the ultimate sovereign deterrence capability.

What is important is to touch on some of the nuclear questions that vex many in the defence domain, because the national security community seems too distracted to undertake useful debate and self-education on this matter, certainly in Europe. The UK is a state that's owned and operated nuclear weapons for nearly 70 years. You could describe it at the forefront of the debate about strategic nuclear weapons from a liberal academic perspective. And that is concerning because in every aspect of the necessary discussion, from the politics and potential military use for them to the exercises of scenarios in which a nuclear guarantee might be expected, there appears to be a deep lack of understanding about when they could/might/should be used, and covering a lack of parliamentary scrutiny over nuclear weapons matters. Any scrutiny in fact. This seems utterly inconsistent with ideas of open democracy, and at odds with the debates and practices of other nuclear weapon-owning states. But these conversations are still not happening.

There is an additional aspect of nuclear weapon capability that rarely sees the light of day: their sub-strategic, or perhaps tactical utility. Since 1989, Western politicians and militaries have placed nuclear weapons in a separate strategic bucket. Regarding them for nearly 25 years as the only deterrent in their own right, the previous 50 years of understanding them as a continuum in the spectrum of conflict and deterrence was forgotten. Increasing liberal mindsets and discussions distracted from the reality of war and warfare elevated ballistic nuclear weapons into a new category sitting outside of military forces: a true political weapon. Whilst understandable, given the narratives playing the West during this period, the real failure was a presumption that every other society would regard tactical nuclear weapons in the same way.

After the Cold War, NATO militaries unilaterally retired their tactical nuclear capabilities. Russia and China did not, and neither did rising nuclear powers such as Iran, Israel, Pakistan, India and North Korea. In Russia, for example, tactical nuclear weapons retained a real battlefield utility to combat the mass of a potential Chinese invasion across Russia's southern border. For Pakistan and India, nuclear weapons continued to play a critical role in tactical deterrence.

The Western separation of nuclear weapons into a different category of warfare (one not concerned with their potential military utility) also started to make Western political leaders think differently about war and warfare. If nuclear weapons had become the deterrent to other states, conventional forces lost that capability, at least in the minds of leaders. The psychology of this taxonomy made cuts to conventional forces much easier to rationalise, particularly in Europe.

The topic of nuclear weapons in war continues to be viewed as a specialist one, rarely discussed outside of a small, specialist and inbred community. Because they perpetuate groupthink built on false assumptions about how the world views such capabilities, this boutique sub set of the national security community has done the West - militaries, politicians and society at large - no favours. The discussion on nuclear weapons (strategic and tactical) needs to be recaptured by the broader profession of arms because potential adversaries did not change their perceptions of them in the way some Western societies did. That failure to understand adversaries would be disastrous in a conflict, and in our efforts to leverage the concept of deterrence against potentially hostile actors.

5

Politics Is Everything

The Western intervention in Afghanistan between 2001 and 2022 was not supposed to be difficult. On the face of it, the under-developed state of Afghanistan was in dire need of investment, mentoring, funding, upskilling, and partnering (amongst other things) after the Taliban regime was removed, and a functioning constabulary installed. The first part of the plan (removal of the Taliban and destruction of terrorist training camps in 2002), went well. The subsequent efforts to reimagine Afghanistan as a modern, gender sensitive, multi-ethnic, centralised state, based on democracy, human rights and the rule of law, failed. Dramatically. In undertaking this effort, the cost to the US economy alone is estimated to be $2.3 trillion, and – more importantly – taken the lives of around 243,000 people.[1]

Given the mismatch in technology, investment, training, support, technology and logistics between belligerent parties (NATO and partner nations versus terrorist networks with local tribal affiliation, covertly supported with information and intelligence by agencies in Pakistan),[2] this seems staggering. Western militaries, and the associated foreign policy diplomatic arms, had trained for nearly a millennia to conduct high intensity combat operations against highly sophisticated and technological peer competitors; and win. Yet faced with an adversary fighting on horseback, with second world war rifles, iron sights, and improvised explosive devices,

---

[1] 'Costs of War', Watson Institute of International and Public Affairs, Brown University. https://watson.brown.edu/costsofwar/figures/2021/human-and-budgetary-costs-date-us-war-afghanistan-2001-2022.
[2] Ahmen Aboudouh, "Where do the Taliban get their money and weapons from?", *The Independent*, 1 September 2021. https://www.independent.co.uk/asia/south-asia/taliban-where-weapons-money-funds-b1911655.html.

the combined power of the Western world was unable to achieve even sporadic control of much of the country. Despite an uncontested electronic warfare environment, no cyber challenge, complete air superiority, utter dominance in surveillance capabilities and technology, and more than 102,000 troops on the ground, Western political and military leadership failed to deliver any form of victory or lasting success. Within weeks of Western militaries departing Afghanistan, the Taliban had re-established complete control of the country, emphaticaly supressed free speech, denied all other forms of political activity, and reversed measures to progress women's rights that had occurred over the previous two decades.[3]

This assessment of one of the best funded military interventions of the twenty first century is sobering: it should also be the cause for a detailed examination of the root causes of failure. Some blame might well be attributed to military leaders and personnel, but a disproportionate part of the failure falls on the successive political leaders who set and then pursued unrealistic goals and failed to maintain a genuine interest in the outcome of the campaign. The rhetoric these political leaders used during the campaign was not matched by their intellectual engagement with reality of situation on ground.[4]

What might have been learned through various deep dives in Western capitals in the year that followed the withdrawal from Afghanistan (2021) was all but forgotten when Russia invaded Ukraine again in 2022. The political, military, academic and analytical community shifted their view and yet another Afghan campaign might well become a forgotten aspect of recent history. More worrying is that the lessons that could have been learned (political, military, diplomatic, economic, and in terms of strategy and policymaking) seem unlikely to emerge or stimulate corrective action.

In an examination of the failures of Western policy during the Afghan intervention, the following discussion is worth reflecting on. It took place in August 2021, just as the final Western military withdrawal from Afghanistan had taken place, and the Taliban regained power in days. That country was, therefore, in the headlines again but at the time it increasingly

---

[3] Andrew Watkins, 'One Year Later: Taliban Reprise Repressive Rule, but Struggle to Build a State', United States Institute for Peace, 17 August 2022. https://www.usip.org/publications/2022/08/one-year-later-taliban-reprise-repressive-rule-struggle-build-state.
[4] Douglas Kellner, 'Bushspeak and the Politics of Lying: Presidential Rhetoric in the 'War on Terror." *Presidential Studies Quarterly*, vol. 37, no. 4, 2007, 622–45. http://www.jstor.org/stable/27552281.

felt that Persia, Arabia, and associated parts of Asia were drifting further from the minds of political and military leaders as their focus shifted to the Asia Pacific. Russia was nowhere in the political discourse of the time – happening some six months before Moscow's 'Special Operation': the invasion of Ukraine. Whether this disconnect between reality and political ambition was a function of political tasking, or the challenge of China and the PLA across the globe, or just wanting to forget the experiences of two unsuccessful interventions over a period of nearly twenty years, is less clear. What was true, however, was that the ideas and balance of grand strategy, domestic politics, and international relations had rarely been more important but also less understood, practiced or developed. Interweaving those aspects of state strategy without a detailed knowledge of the lived experience of the world and conflict is, as it was in August 2021, fraught with the danger of yet more uninformed missions and unrealistic expectations. There is an ongoing requirement, an obligation perhaps, to tackle such a topic. But finding the right guest to do so was challenging.

There were few people who had the requisite experience as scholar, politician, soldier, diplomat, traveller. Rory Stewart was someone who could rightly claim all those titles. As a traveller Rory walked across Asia, eighteen months yomping across Iran, Pakistan, the Himalayas, a solo walk across Afghanistan, as well as shorter treks across Western New Guinea and covering much of the United Kingdom.

As a politician, he served in the British Cabinet and on the National Security Council in various ministerial positions, eventually resigning and returning to Yale to teach politics, grand strategy and international relations. Rory did a short commission in the Black Watch (an infantry regiment of the British Army), was rumoured to be a spy for Britain's Secret Intelligence Service and worked with various military formations during his periods as diplomat in Iraq, notably during the siege of his compound in Nasiriyah Iraq in 2004. Rory moved to Kabul in Afghanistan for three years to run a human development NGO to restore the old city and then relocated to Jordan to work near the Golan Heights on a project to restore a Roman site. Whilst born in British Hong Kong, Rory lived all over the world. Lived in its truest sense, with the people walking their roads, pathways and living their existence. Whether that is as a soldier, a farmer, a trader, or a fellow traveller. His detailed knowledge of Afghanistan, as well as of politics and diplomacy, and certain familiarity with military matters made him the ideal commentator on the topic of contemporary politics and war.

What Does the Western Way of War Mean to You?

Rory Stewart: It's a great question. For me, I think at its best and this, of course, is me talking about the kind of warfare that I admire and I am interested in, is about something that is genuinely attuned to politics that is thoughtful about history and context that works with communities. Of course, that's what has always seemed to me to be distinctive about things that have worked well in Bosnia, or Sierra Leone and even those things that seem to work best in Afghanistan or Iraq. I would like to see a way of war that integrates it much more powerfully with intelligence agencies, diplomats, aid agencies and soldiers and really make that work. The missing key to that is politics; usually people talk about integrated approaches but what they fail to take into account is politics in the sense of local power, really understanding how local power works and how to work through that.

    Peter Roberts: When you say, "At its best", that's what you'd like it to be, but you've experienced quite a lot through your career in various ways, whether as a scholar, or as a traveller or as a soldier or as a politician, as diplomat, whatever that is. Actually, we often fall short of that 'best' right, and Afghanistan is one of those areas where we have perhaps fallen short, would you agree with that?

    RS: Yes. At its worst, the Western way of war, is predicated on various [misconceptions], a sort of almost mental illness. It's incredibly binary. At its worst, it believes every problem has a solution. It tries to fling astonishing amounts of money and troops at trying to overwhelm problems and lurches from a surge [of resources] to total withdrawal and is fuelled by paranoia and megalomania. If I just bring that down to earth in Afghanistan, if you look back at the way we talked about Afghanistan at our very worst which was 2007 or 2008, the paranoia was saying, 'Afghanistan is an existential threat to global security and if Afghanistan falls, Pakistan will fall, mad mullahs will get their hands on nuclear weapons.' The megalomania was, 'And we are going to fix this, we are going to turn this country into a gender sensitive, multi-ethnic, centralised state based on democracy human rights, the rule of law.' The great catchphrases which really reveal the madness of this particular Western way of war, was the idea that failure was not an option. When failure was all around us, we kept reiterating failure is not an option, and every twelve months we would again have a senior figure saying, "This is the decisive year", whether that was in 2006, 2007, 2008,

'09, '10, or '11. I have all the Generals on records saying their year is the decisive year.

PR: If we have a set of unrealistic approaches then that's not just politically, that's also militarily: it's the Generals saying this. But is that just simply because of a failure to appreciate local power, is it megalomania, or is it something to do with domestic politics, both at home and abroad. It's just not a failure to understand the power relationships in Afghanistan, but also a failure to really be truthful about what we are trying to achieve at home, correct?

RS: Absolutely. Failure to be truthful in every single way. Now of course, politicians quite rightly get a lot of the blame for this, and they are often the villains of this piece. But it is surprising how often the Generals are also to blame, and how often the analysts are to blame. How are they all to blame? They are all to blame because they're perpetrating different types of untruths. The politicians are saying things which aren't true because they want to sound generally positive. The Generals are saying things that are untrue because they think it's vital for the morale of their soldiers, to claim that they're winning, and that it's all going well when its patently obvious that it isn't. The analysts have an incentive to tell untruths because they're employed not to point out problems but to present solutions, so the analysts are, sort of, snake oil salesmen who are encouraged to turn up, paid a thousand dollars a day to tell the politicians, the Generals, that they have the magic bullet solution to fix these things. These are all different types of lies. What nobody was prepared to say in Afghanistan is that with a great deal of patience and a great deal of luck, we may be able to make Afghanistan over 20 or 30 years feel a bit more like Pakistan and a bit less like the Congo. What we had was a series of people claiming to be realistic but actually producing nonsense, endless different forms of nonsense and I realised there were two things going on there. The first thing of course is that most of these people knew nothing about Afghanistan so even if they were based on the ground, they didn't speak a word of an Afghan language and security conditions meant they were in locked and guarded compounds.

If they interacted with Afghans at all it was in a very, very odd way, through translators who are often not from the region they were in so they simply didn't know very much. But secondly that the entire incentives of the military, the politicians, the diplomats, the aid workers, and the analysts was to placate an audience, which could be the audience of the soldiers or it could be the voters at home. None of them had an incentive to try to accurately describe what the conditions were like in Afghanistan, and in fact they would have found it almost unbearable to try to provide an accurate description of what was going on, because it would have seemed impossible.

PR: Now it strikes me that there were so many failures, at so many levels, whilst we were dishonest with ourselves perhaps for thinking about how we could reshape a country to be something that was an unrealistic proposition. As a coalition we didn't shirk from throwing money, people, studies, good will at it. So, it can't simply have been just a case of money, materials and military force that failed, to enable us to make the country a slightly better place. Was it just our appreciation of power, was it local knowledge, what sits at the heart of this, if we took out the sort of megalomania, and the strangeness of policy at the top, is it simply a failure in grand strategy?

RS: It's that Afghanistan itself consisted of twenty thousand villages, it was one of the poorest countries in the world, you were recruiting in Helmand a police force in 2011, where 92 per cent of your recruits were illiterate. Where there was no electricity between Herat and Kabul. Where the ethnic tensions between the different groups was intense, where there was support from the Taliban coming across the border from Pakistan, where there was no real functioning state. Those things could never be overcome, it didn't matter how much money you spent or how many troops you put on the ground. Working with the grain of Afghan society, required what we were doing before 2005/06. The period from 2001 to 2005/06 when essentially the Afghan government had very little support from outside actors and ran its own affairs was infinitely preferable to the situation that followed when the West tried to surge [resources]. First, the British went into Helmand and General McChrystal got us up over 1,000 troops. The best thing that happened to Afghanistan was Iraq in 2002/03, because it distracted the politicians and soldiers away from Afghanistan and kept the footprint in Afghanistan light. Forced the Afghan government to take the lead. Now there were many things that were extremely disturbing about that period, 2001 to 2005/06, which is why people supported the surge. There was, of course, endemic corruption. Many of these provinces were ran by corrupt warlords. There was a strong Taliban presence in rural areas. But we needed to recognise that we had no better alternative, that the idea that you could somehow replace power holders in Helmand or Herat with technocrats from Kabul, and try to run a centralised state, backed by foreign military forces and that would somehow be more effective, was madness. It was incredibly costly in terms of lives and money and ultimately made the whole thing feel unmistakably – to the Afghans – as though they were living under foreign military occupation.

PR: Which they pretty much were; there was absolutely no way you could doubt that with the post surge levels or surge levels of troop commitment that was there. From everything that you are saying, it just strikes me that surely we can't have gone into this thinking that Afghanistan was like another western state with a centralised government that ran everything in a sophisticated infrastructure and the central policies that would be followed by the 20,000 villages. It feels like we just didn't understand the country at all and yet I find that quite hard to believe. The UK and FCO (Foreign and Commonwealth Office), now the FCDO (Foreign Commonwealth and Development Office), has a long and illustrious history, not just in Afghanistan but of understanding local communities. Did we get it so badly wrong in our initial assessments about the country, is that what lies at the heart of our problems with western intervention?

RS: We got it unbelievably wrong and remember that British diplomats are not people who are experts on the rural areas of countries. They operate from embassies, they deal with ministers, they live in capital cities. We are a long way away from the days of the British empire where you had district offices out in remote rural areas. DFID (UK Department for International Development) staff are increasingly very, very few people managing enormous budgets out of central capitals. They are really project management people who administer a £50 million or £100 million budget and, again, they are not people who understand the rural areas of Afghanistan well. Their targets from the start are totally unrealistic. Let me give you a tiny example, which sounds as if it's a long way away, but I'll try to link it. DFID in Afghanistan did a project on carpentry and I was running a carpentry shop in Kabul. Running the best carpentry school in Kabul, we were training some very fine carpenters over three years with a lot of investment. I went to DFID and I said, 'Terrific, you have got a £50 million program for carpentry training, can we have some of the money?' 'No, no, because you are not training enough people, we want to train 10,000 people and we are going to put them through a two-week course.' I said, 'Well, look, this is mad. Nobody can learn carpentry in two weeks' and they said, 'Yes, but if it is longer than a two-week course, we can't train 10,000 people.'

When I questioned it; I mean obviously it's a bright person I am talking to, he, sort of, gets the problem, but he feels part of the bureaucratic system, which has trapped them. Another example from DFID in 2004/05 when I said what we are doing is nonsense, they said, 'Oh no, it's not nonsense, what we are doing is we are paying teacher's salaries.' Who can possibly disagree with that? An Afghanistan NGO (Non-Government Organisation) then did a study in Ghor province in central Afghanistan and found that the three thousand teachers being paid by DFID didn't exist, they simply weren't there.

They were just names collecting money and they were nowhere near a school. The fact is that our civilians (as much as our soldiers) could not begin to take on board how extreme the situation was. They mirror image [their own experiences]. Of course they understand it's a poor country but they haven't spent enough time in the rural areas to just start by saying, 'Is it likely that we can train 10,000 carpenters in Helmut? Is it likely that there are three thousand teachers in Ghor province receiving a salary?' No, so something must be wrong here. For example here, General McMasters, moving onto the military, he arrived as a sort of one star [officer] and his job was to eliminate corruption in Afghanistan. I remember the conversation there where I essentially said, 'You can't eliminate corruption in Afghanistan, nobody has managed to do this in almost any country in the world. They haven't eliminated corruption in India, they haven't eliminated corruption in China, and those are much, much wealthier, more developed countries, they haven't eliminated corruption in Saudi Arabia, what do you mean you are eliminating corruption in Afghanistan?' To which the military answer was, 'Well, okay, don't get me problems, give me a solution, how many troops do I need, how much money do I need? I'm going to crack this'. He ended up setting up divisions of soldiers that were going to eliminate corruption. But it's insane, right? Somebody needed to say, 'This whole thing is insane, you are a one star general, you cannot eliminate corruption, stop asking me how many solders you need, how much money you need, you can't do this. You don't even understand what you are talking about. What are you imagining you're going to do?'

PR: It's incredible that we need that kind of [intellectual] challenge in many ways, because the problem – it seems – is often not with those who have been deployed to the field who have been given a role, they're go getters and their promotion is on the line, they might be in the country even for a couple of years, but they've got to deliver it. Instead, it comes back to systemic problems in the [remote] capitals in the West. From what you're saying, that says we are driven by completely the wrong set of targets and metrics, that we think of deliverables in our own terms and not from the others [perspective]. It feels like you're saying that we can't somehow appreciate the context, we must make everything about ourselves, and how we would like it to run, rather than the realities. Are we just not prepared to engage with the realities of situations on the ground?

RS: You're right, we can't. I used to think this was the only problem but then I heard the President of Afghanistan say every Afghan is committed to a gender sensitive, multi-ethnic, centralised state, based on democracy, human rights and real law. I literally cannot translate this into words that anyone in an Afghan village can understand; I don't know how to say that in Dari and obviously it's not true. I've just been walking for 32 days through villages which were at war with each other and when I haven't seen a woman for 32 days; what do you mean they are committed to being gender sensitive? I thought, 'Is this guy cynical or is he naïve? Is he lying or is he ignorant?' and of course, it's not just him saying that, Gordon Brown is saying something similar, David Cameron will say similar things, Tony Blair will say similar things, George Bush will say similar things and the Generals will say similar things – maybe not quite as extreme but similar sorts of things about how they're going to sort it out. But what I realised is that it's not that they are either lying or ignorant, it's that they're not really talking about Afghanistan at all. [To all of these people] they would actually be a bit sort of shocked if I said, 'Wait a second this isn't a description of Afghanistan,' because what they're really doing is talking to some other audience, an audience at home.

If I said to them, 'Okay, try to describe Afghanistan': this is a country which is on the edge of civil war where the Taliban have enormous control over rural areas where in most of central Afghanistan women don't leave their houses, where the literacy rates are 8 per cent, where there's no electricity, where there's no allegiance to the central state and [now] we are going to try over 20 or 30 years to make it a little more like Pakistan, a little less like the Congo. They would say, 'Well, we can't say that, Rory. It doesn't matter whether it's true or false. You simply can't.' I mean, how are you supposed to get the British public to put money and troops into that project? Let's take South Sudan which I've been thinking about as well. If you were to say in South Sudan, 'Look, we're not going to be able to create a functioning state. We might be able to get some humanitarian aid in to feed some starving children if we accept the fact that the militia's going to steal some of the food aid on the way in and we'll be able to keep a few people alive.' No politician is going to want to say that to their public. Nobody's going to put £100 million a year of tax-payers money behind that. They have to keep – or they feel they have to keep – pretending that it's all going swimmingly. That there's a great sunlit future. That there's some great hope that they can offer and that's in the structure of the way that the West thinks. It simply doesn't have the patience or the tolerance for saying, 'This is a deeply traditional society that's very challenged and we're going to very slowly work to try to improve things.'

That then leads me to my real sadness which is that this tendency leads them from a surge (a massive over deployment of troops), to total withdrawal because they have attention deficit disorder. They go from total over-optimism to complete pessimism because they have no sense of reality. I'm running an NGO on the ground in Kabul. Ironically, I was more optimistic than the people that were spouting all this apparently optimistic nonsense because I actually had a sense of what Afghans were capable of. We were creating real markets. We were selling carpets. We were getting a clinic going, we were getting a school going. I had a real sense that we could get stuff done. What I realised is that the West talked an optimistic game but was basically profoundly pessimistic. The reason why they spun all this nonsense is they, at some level of fear, didn't really believe they could do anything at all. They, sort of, despised Afghans rather than admired them.

PR: Is that because they couldn't make the leap of imagination? Is that because they just didn't have the experience on the ground or is that because they'd been set the unrealistic goals of previous administrations of creating this nirvana? It just feels like it couldn't possibly have happened and yet no one walked back from it. Then the surge became binary. [They must have thought] "The surge didn't work, the whole thing's going to a ball of cheese, get everyone out." The context – again – is just not being explained. A new administration, you'd think (whether it's in the UK, amongst allies, or whether in the US), has the opportunity to break with the past at every change of leadership that has happened and yet there's been none of that.

RS: Yes, it's very strange. [US President] Biden has just made the most reckless damaging and unnecessary decision imaginable. He's pulled [all his forces] out [of Afghanistan] and by doing so essentially handed the country to the Taliban in about two and a half months. It is extraordinary. Yesterday, they managed to take Pol-E Khomri. The entire road from Kabul North has now gone. It's incredible, the advances they've made; district capital after district capital is falling. He [Biden] had absolutely no reason to do this. There have been no [Western] combat operations since 2016, they [the US military coalition] were containing the situation with relatively few troops. They [Western forces] were down at 2500 troops and a few planes. They were not suffering casualties. They could have remained there indefinitely in the way that we do in Germany, Japan, South Korea. Instead of which he [Biden] asked this completely daft question which is, 'Well, when are we going to leave? and surely if we leave in 10 or 20 years time the same things going to happen,' is, sort of, irrelevant. If it's not costing you an enormous amount and by remaining you are stopping the Taliban running over the country and remember that is stopping horror. It's stopping millions of refugees flooding out of the country. It's stopping suffering of millions of Afghans. It's stopping the total humiliation of the United States and its Allies, and potentially it's also stopping the re-emergence of a terrorist safe haven. It's a pretty low-cost investment. What he has done is so beyond imagining reckless and irresponsible and such a betrayal of our Afghan partners. Nobody there is responsibly and slowly trying to work out what we are going to do to unravel our programs.

There's all this nonsense. 'We'll take the military out. We'll keep the aid programs going.' It's complete nonsense. The entire insurance system in Afghanistan has now collapsed, insurance companies have all withdrawn. How do you keep the aid programs going when all the foreign powers have just told all their foreign nationals they have to leave the country? All the civilians have been instructed to leave the country. The US government has announced a visa program which will mean hundreds of thousands of Afghans, almost every educated English-speaking Afghan is going to be trying to leave the country now and nobody has thought patiently about what do we do to try to manage a transition. They've literally just swept the rug out of the things, washed their hands of it and are pretending it's not their fault. I mean, it's extraordinary. Absolutely extraordinary. I mean, in human terms, it's as though you'd decided to foster someone into your home, been working with them, developed their trust, developed a relationship over 10 years, and then suddenly just kicked them out the house, locked the door on them and set up nothing for them to do. Can you imagine the sense of disillusionment and rage from Afghans at this sort of treatment?

PR: It's also pretty embarrassing that Western capitals should behave like this. From everything you're saying, and we would all recognised so much of it, we might feel perhaps that we have a civil service that's run by targeting catchphrases, that's agnostic of the realities of the ground of the context in which they're operating. I am certain that politicians will have been shocked to realise this, but the public surely must be shocked as well. Shouldn't we be expecting more depth, more candour, more reality, more detail, more expertise from our civil servants? Should we not be expecting the FCDO to understand this context; to be explaining the impacts of the political decisions that they're making? Or do you think this is going on and that the politicians are just ignoring it?

RS: Well, one of the problems is the Foreign Office began to be cut so heavily from 1997 onwards. The core central diplomatic stuff is about half the size of its French equivalent with an economy about the same size as France. Its language training was cut to pieces. It became very, very risk adverse. You know when I was in Foreign Office we had 26 British diplomats in Zambia. By the time I was Africa Minister [2017] we had two, which was the ambassador and an assistant. The incentive structures for promotion in the Foreign Office have nothing to do with developing deep expertise or learning languages. It's all about performing in Whitehall, demonstrating your management capacity. It's become very inward-focussed. People are rewarded for good corporate approaches, showing sensitive approaches to managing their own teams. They're not particularly rewarded for getting out of the embassy, spending hours developing contacts with foreign nationals or learning their languages and that's been going on now for 30 years so that somebody like me, if you'd go back to when I first went to Afghanistan, the end of the Taliban period and I was on leave from the Foreign Office so obviously I returned to the Foreign Office after my walk and started saying, 'I think you've gone mad. This is what I think Afghan villagers are like,' and the problem is the culture of the foreign office was incredibly dismissive. Their sense was what I was saying was uncomfortable. What I was saying is that a lot of what you're claiming you can do you can't do and a lot of what you're trying to do is simply impossible and it's going to waste lives.

Their response was not to try to engage seriously with what I was saying but to attack me and other people like me and say that we were fantasists, we were romantic, we thought we were Lawrence of Arabia, we were living in the 19th Century, we didn't get the modern world. Or if that didn't work, they'd say, 'Oh, you know, he's a bit unreliable. He's a narcissist. He's self-promoting. He's only doing this to get media attention.' It became quite extreme. I mean when [Sir] Sherard Cowper-Coles, who was one of the most senior diplomats in the British government and was the [UK] Ambassador to Kabul, tried to say that the counter insurgency strategy was madness, they did the same to him. They effectively fired him, deprived him of his next posting and made him out to be a mad man. The same happened to General Eikenberry. He came in as the US ambassador. Very distinguished American General. Deep expertise in Afghanistan. Three distinguished tours in Afghanistan before he become US Ambassador. As soon as he started saying the surge was a mistake, everybody said he was senile. He'd lost the plot. He had no idea what he was doing. The same actually happened to [President] Biden in the days before he did his current madness. In 2008, he was the only person arguing for light long-term footprint and the response from the national security establishment was to suggest that Biden was senile, inept, stupid and they never engaged seriously with this argument that actually we could've had a light footprint and relied more on air power.

PR: All those things you say I recognise across militaries as well as the FCO. The obsession with promotion based on budgetary performance, on narratives, on relationship with technology, the faddism of cyber open-source intelligence, the human element has all but disappeared. The acceptance of mavericks, the understanding that deep expertise has enormous value disappeared in the 1990s and – as a culture – has gone entirely. But what do you think we would need to do? Because I think we must want to be able to recapture a seriousness, a gravitas in our civil service, right across Whitehall, right across the military as well. What might we do to start re-instilling some of those good practices right across Whitehall that allow us to get back to the best of the Western way of war as you described it at the start?

RS: I think one thing is that we can learn from the best of the United States. The Americans have just done the most terrible thing. [President] Biden has just done the most terrible thing but at their best American politicians, American generals, American diplomats, in my experience in Afghanistan, were far more serious, far better informed [than their British counterparts]. [The Americans] Took much more of a sense of responsibility. They were really prepared to spend the days going through the detail of the challenge, and the counter-insurgency warfare doctrine. They were really prepared to invest in the anthropologists. They were really prepared to rethink their tactics. They were very open to outside advice. I remember back in 2008 I was spending time with John Kerry and being mesmerised by his willingness to talk about North Waziristan, South Waziristan, associated federations – stuff that no British politician could do or was interested in doing. I was very impressed. Even though I disagreed with what they did with Generals Petraeus and McChrystal. Right, they were, seemed to me, at a totally different level and so I think we can learn from the United States. I think right away across the board we can learn from the [US] Rangers and the US Marine Corps as good expeditionary formations. We can learn from the way that they send some of their best people off to universities. Their training is much better, but particularly we can learn from the fact that they have the budgets and the openness to open up to outside experts to take that challenge. Take [US General] Stanley McChrystal for example. As soon as he got in country [Afghanistan], he gathered people around him who profoundly disagreed with his point of view and spent a lot of time listening to them; and he wasn't doing it proforma.

I feel when the British civil servants or politicians do it, they're just going through the motions. They will endure being forced to sit there with a critic and they'll be polite and then leave and not think about it again. Stanley McChrystal was wrestling with it. He really wanted to know whether he was right or wrong because he felt that he was carrying a responsibility for it in a way the British didn't. I would like us to be training up people. I would have liked to take [UK General Sir] Nick Carter's developments far further. I'd like to really lean into the idea of foreign area expertise but *really* lean into it and that means having the budgets to make sure that people can learn languages, can travel, can get out [of London]. One of the embarrassments as we increase the Ministry of Defence's budget is that there still never seems to be any money for deployment. For example, when I wanted to work with the military in Africa, it was a perpetual problem getting a few £100,000 together to get people on planes. So these structures would be created [to try to make that happen]. The idea was meant to be there but there was just no flexibility to actually make anyone do anything. We'd end up with these embarrassments of tiny, sort of, £9 million operations which we claimed were going to defeat Boko Haram or 250 engineers sitting in the inner circle of a UN base in South Sudan pretending that they were doing peace keeping. Yes – there's a lot we could do and it probably begins by learning from the best of the Americans.

PR: As a final question, I wonder if I could ask you if people were thinking about trying to get serious about context, about future conflict, about the Western way of war, about integrating approaches, what advice will you give them right at the start of their career? What should they be looking at? Is it a case of reading more? Is it a case of travelling? Is it a case of walking? What is the secret to us developing this group of people who will go out and reinvigorate our seriousness?

RS: I would think that the real secret is to make sure that people, after they've been in four/five years as an officer and they have talent in this direction, I would give them unpaid leave to do very tough, extreme travel in remote rural areas. There's simply no substitute. For example, sitting in a village in central Afghanistan or it could even be spending five/six weeks in a village in Western Nepal, changes everything, because just living that life with those people makes you understand who they are. And from then onwards for the whole of the rest of your career, when somebody says, 'Could we have a Facebook revolution in Western Nepal?' you think, 'Well, okay, let me think about this. We didn't have any electricity. Nobody was literate. How are we going to do a Facebook revolution, right?' So, I think that is the absolute key and it's not just the key for the military or the diplomats, but the development workers. The development workers, I really believe every single DFID staff member deploying needs to spend time in language training, needs to spend two months in a remote village before they start their job in that country. Because if they don't know what a remote village in Kenya is like they can't run the program out of clarity. But nobody will let you do that. They never provide the budgets. That tiny investment seems to be beyond people because they pretend they live in a world in which everybody speaks English, everything can be done from a [nation's] capital, and everything can be done remotely. Increasingly, they question whether they need anyone in the field at all because it's much cheaper to base them in East Kilbride and hope they can run the Kenyan program from there.

But I think that would apply actually to almost every Westerner, whatever they're doing – one of the things that's missing with so many young people; it doesn't matter where they're going into business or they're becoming an academic or they're just thinking about what they want to do with their life – is any exposure to how people live in poorer, remoter areas of the world.

Reflections

States that have had empires – physical, economic, or ideological – have historically developed a good understanding of local and regional dynamics in countries where they have had some kind of national interest. A civil and diplomatic staff able to interpret and assist in the development of those states towards a future which is mutually beneficial. There is also a peculiar moment for declining powers – whether economic, militarily, or simply in terms of outlook and influence – in which the cumulative knowledge banked over generations of diplomats becomes unfashionable and is, perhaps, viewed internally as a burden rather than as an asset. This might sometimes be associated with a period of economic frugality or with a political vision that disagrees with the assessments of those working in a semi-independent civil support function.

At this point, what often follows is a significant 'change programme', designed to remove expertise that does not suit the political agenda. It usually entails cutting deep experts from staff with the associated loss of knowledge that is difficult to regain. Whilst such expertise, delivering knowledge and understanding of local and regional issues, can be regained provided that time, effort, resource and interest are delivered by political leaders, what is more problematic is when the culture of an organisation changes and removes the 'seriousness' (as described by Rory Stewart above), from institutions: at that point the knowledge, understanding, and the processes to influence and make an impact built over generations, as well as an understanding over how to make decisions in a manner cognisant of their fullest implications is irretrievably lost. This is as true for military organisations as it is of civil service ones.

There is a moment at which political leaders fail to see the utility of both diplomatic and military levers of power, seeming to believe that industrial might, trade, technology, and their own political 'wisdom' has no need for the detailed understanding and advice that professional and semi-independent ministries can offer. That point of political hubris marks the moment when state power fails entirely. The Western intervention in Afghanistan between 2001 and 2022 is a case study in such failures. We should spend more time reflecting on these failures.

6

Hybrid Warfare

Since Frank Hoffman and Jim Mattis originally coined the term 'Hybrid Warfare' in 2007,[1] it has developed an academic school of theory around it. Moving so far away from the original concept so as to be unrecognisable from the original, Western militaries – perhaps looking for a new theory of warfare in a world where it was perceived that high intensity, conventional combat would no longer take place – embraced the ideas around it, fusing some of the more rigorous thinking with a thin-slicing[2] of technological promise from a new generation of science fiction writers. The theory itself developed a new language, developing a taxonomy of sub-liminal, threshold, and grey zone warfare that was to become so fashionable between 2014 and 2022. Even as Western intervention in Afghanistan was drawing, unsuccessfully, to a conclusion, military theorists were shifting away from doctrines of Counter Insurgency and Counter Terrorist operations to one of Grey Zone warfare.

There was such a deep belief in the minds of Western political and military leaders that this 'new' type of warfare would dominate military activities for the coming decades that every incident, event and activity across the world became painted as grey zone activity. Western analysis of the Russian invasion of Ukraine in 2014, their annexation of the Crimea peninsula and occupation, was so obsessed with this mantra that it attributed this Russian operation as 'hybrid'. It became the archetypical example of this type of warfare, despite the Russian use of massed conventional

---

[1]   Frank Hoffman, 'Conflict in the 21st Century: The Rise of Hybrid Wars', Potomac Institute for Policy Studies (Arlington, Virginia), December 2007, https://www.potomacinstitute.org/images/stories/publications/potomac_hybridwar_0108.pdf.
[2]   See Malcolm Gladwell, *Blink: The Power of Thinking Without Thinking* (Black Bay Books, Little Brown, 2005).

military as the primary arm in this campaign. Little Green Men became popularised in Western media as emblematic of a new Russian doctrine where hybrid sat at the core.[3]

Indeed, at the time Western analysts claimed that this military doctrine was the child of General Valery Gerasimov, then Chief of the Russian Army, after misreading a speech he gave on activities in the grey zone. While Western media and political leaders claimed Gerasimov was outlining a new way of fighting for the Russian military, what he was describing was how they needed to understand the Arab Spring revolutions and the way in which Russia could protect itself against such activities.[4] But the truth was immaterial to Western leaders. The Gerasimov doctrine began to spur an entire industry, including podcasts and television shows. Western militaries took note – believing they had discovered a new threat matrix of adversaries and began to reorganise around that threat. As noted by Vladimir Rauta and Sean Monaghan, "The United Kingdom's integrated defense and security review [2021] put "grey zone" or "hybrid" challenges at the center of national security and defense strategy. The United Kingdom is not alone: The security and defense policies of NATO, the European Union, and several other countries (including the United States, France, Germany, and Australia) have taken a hybrid-turn in recent years."[5]

The idea that hybrid warfare was to become the predominant and sole way that adversaries would wage war was never a compelling argument for 'transformational' change of military structures, force design, decision-processes, procurement and training. But such arguments were not welcomed by senior political and military leaders in 2021. A more considered approach – acknowledging grey zone activity as just one of many approaches that adversaries could use – was only to be found in small rooms, well away from those obsessing with a supposed new way of warfare. The following discussion is worth recounting as a more reasonable

---

[3]   John R. Haines, 'How, Why and Where Russia will deploy Little Green Men – And Why the US Cannot', Foreign Policy Research Institute, 9 March 2016, https://www.fpri.org/article/2016/03/how-why-and-when-russia-will-deploy-little-green-men-and-why-the-us-cannot/.

[4]   Mark Galeotti, 'I'm Sorry for Creating the 'Gerasimov Doctrine'', Foreign Policy Magazine, 5 March 2018. https://foreignpolicy.com/2018/03/05/im-sorry-for-creating-the-gerasimov-doctrine/.

[5]   Vladimir Rauta and Sean Monaghan, "Global Britain in the grey zone: Between stagecraft and statecraft", Contemporary Security Policy, vol. 42, no. 4 (2021), 475-497.

discussion over where and when the West got confused about hybrid/grey zone activity.

Ewan Lawson was a senior research fellow at RUSI in London who spent a good deal of time thinking and talking about hybrid/grey zone warfare between 2015 and 2020. Prior to joining the think-tank community, Ewan spent nearly 30 years in the British military working across the joint force well before it became fashionable. His wealth of experience, from commanding 15 PsyOps Group to setting requirements for the UK's National Offensive Cyber Programme, as well as working across government at the UK's Permanent Joint Headquarters, deployed to the world's hotspots, or as Defence Attache to South Sudan, made him one of the foremost soldier-scholars on the topic. He spent a lot of time talking to authors of these theories and debating their relative merits with distinguished academics and practitioners. I disagreed with Ewan quite often, which always made our discussions so useful and worthwhile, but his pragmatic approach and realism to the challenges being faced in competing between states below the threshold of open military hostilities was instrumental in providing some gravitas to the 'new' theories of war.

What Does the Western Way of War Mean to You?

Ewan Lawson: I'd say the Western way of warfare as we now conceive it – and I think it's not static, it changes over time – is one that is about the overwhelming use of force. It is increasingly about the use of precision and it's increasingly about risk – perhaps perversely, because whilst the overwhelming use of force is at the heart of it, at the same time it is trying to do that with zero or certainly minimal civilian casualties (which is an entirely appropriate aspiration if perhaps challenging in certain contexts).

Peter Roberts: Do you think the Western way of warfare will continue to evolve in this way such that the drive towards zero, minimal casualties will change this idea of overwhelming use of force as a central facet?

EL: I think what we're looking at here is a fairly recent development as you rightly say. You're absolutely right. If you look back at the Second World War, the whole philosophy of 'metal-not-flesh' was at the heart of the allies' approach. The overwhelming use of as much equipment, as much capability as possible, and therefore minimising the risk to our own soldiers at that point. Personally, I point the finger – and I've been criticised for doing this – at the Liberation of Kuwait as being a critical turning point in that whether the reality which was precision, precision, precision isn't really the point. The point is the narrative was precision, precision, precision. The fact that actually a relatively small proportion of the weapons employed in the Liberation of Kuwait were precision munitions is forgotten. It's all about how the overwhelming use of force in precise ways enabled a quick victory and therefore minimum casualties, albeit one only has to look at the Basra Road to see that actually that's not the reality. But, it is the narrative and I think we've rather caught ourselves up in that through Kosovo, through the air campaign, and indeed through pretty much all of the subsequent conflicts. It's this overwhelming focus on precision and this is one of the things that our adversaries, of course, are deliberately trying to use against us in a perfectly logical, strategic approach.

PR: Is it correct to say that in Bosnia, after the UN mission failed and NATO intervened, the use of precision munitions to achieve a success that didn't come with zero civilian casualties, was seen as the lesser of two evils?

EL: Yes, I think I would question completely what you're saying there because I think if you look at the employment of NATO in Bosnia, that's much more about the use of boots on the ground with robust rules of engagement. By the time we get to Kosovo, the narrative is much more about precision. In part, that is successful. You can debate the reasons for the ultimate, we could say capitulation, but the ultimate standing back of the Serbian Government: My own personal view for a long time has been much more about the pressure that was being put on the finances of those who were supporting and surrounding Milošević. Some of that involved the targeting of industrial facilities with links to them and the threat of that which you can do with precision and I think again it's about imagery. What you didn't see was a Belgrade that looks like Berlin in 1945. Of course, the counter to that is, you saw a Mosul not that long ago that did look a bit like a Berlin in 1945, if not worse, which I think rather puts a lie to some of this.

PR: When we were using airpower weapons of precision, and it was about boots on the ground, [from the end of the Cold War] right through to Kosovo, this became more about the narrative and less about the reality and more about how you sowed this. Did this count for home audiences or did it count for adversaries as well?

EL: I think what adversaries spotted was that home audiences were increasingly expecting the reduction in civilian casualties and, of course, when you talk about those campaigns, it's sometimes really difficult to know. Civilian casualties recorded which you could probably do a whole episode about this on with people much more competent on the topic than me. Civilian casualty recording is a real issue because we don't actually know half the time. Unhelpful, I have to say, was the British narrative that there had only been one or two civilian casualties as a result of British air strikes. That's actually not what we said. What was actually said was that there was only one or two *confirmed* civilian casualties as a result of British airstrikes but it was an unhelpful way of saying it for me because that's probably not the reality. But civilian casualty recording is challenging. I think it is about a narrative. It is about a narrative which, at one point, was probably very useful for the military because it was then about the ability to procure equipment, advanced equipment, precision equipment, and why wouldn't we want to do that, if that minimises the risk of excessive casualties? But I think our adversaries looked at this and said, 'Well, this is one of the areas in which we can now potentially exploit this reluctance for our own casualties as well as our reluctance for civilian casualties.'

PR: Would you agree that, as adversaries have started to use narratives which have a long history of success, these narratives are being waged and used better by such adversaries than they by Western powers?

EL: I think the thing that has changed is the digital revolution and I think, for me, however we end up conceptualising, defining, hybrid, grey zone, whatever you want to call it this week, the thing that is really different (if anything is different at all), is the impact of the digital revolution. I was thinking about this the other day in terms of saying where people got their news in the UK 20 years ago, half a dozen mainstream newspapers and a couple of television channels. Now, you have a proliferation of television channels. But, probably more important, the proliferation of social media as a relatively uncontrolled space. Indeed, a space which at various times the regular news media have exploited. It's not uncommon to see mainstream news outlets put at the bottom of a story, 'Were you there? Did you see anything? Send us a message,' because journalists have been cut back in numbers and the funding and the support for journalism has declined, relatively speaking, I think. So, I think this is for me the thing that almost defines this challenge and I think some of our adversaries, partly because if you're a regime concerned about regime security, the Internet (and social media more specifically) represent a particular challenge to you and a particular threat. Perhaps in looking to defend themselves from the threat from social media, they also spotted the opportunities to undermine Western states. I do think the key thing here is – indeed, one of the real challenges of this whole conversation is – where is the boundary between traditional statecraft but just delivered in the digital era and labelling pretty much everything as warfare. That for me has been one of the most concerning things: labelling everything as warfare tends to point to the military as a solution, whereas actually a lot of this is more about educating young people to engage critically with material they find on the Internet.

PR: This idea that it's the military's problem is one of the most important facets of hybrid warfare. I think that came out of 2014 in Ukraine and the West being caught by surprise by something it hadn't seen coming. The popular narrative saw a new type of warfare that employed little green men and narratives and cyber to annex whole regions. It strikes me that this was the start of the hybrid industry, would you agree?

EL: I think 2014 is a significant turning point and is part of the problem. I think you can make a case that one of the reasons there was a sudden spike in this was the sense that we, in the West, have been caught out here. So, it clearly can't be that we got things wrong, it must be that there is something new happening here. I think that we did drive something of a hybrid narrative which has moved quite a long way from Frank Hoffman's original concept of the blurring of conventional and irregular warfare with terrorism and a bit of crime and so on and so forth which, of course, is also a bit like Mary Kaldor's, 'New Wars.' Again, the problem with both of those is – not about problems, I think it's always worth reminding ourselves about these things is – that again those sort of things have been going on throughout history. So, 2014 is significant, but 2014 I think is both significant in terms of the turning point but it's also the point at where we really start to get confused about what it is we're talking about.

So, Crimea, stand-alone Crimea, is a military operation fundamentally with a significant element of obfuscation designed to slow down the political-military response of the West if there was going to be one. Just by putting enough doubt in there, politicians were saying, 'Perhaps we aren't seeing what we think we're seeing.' There's just enough doubt. The problem is that label then was applied to pretty much everything Russia has done subsequently, and this is why I start to have an issue because there is no doubt, in my mind, that Russia is trying to keep Ukraine destabilised and it is doing that in part through its occupation of Crimea and also in parts of the East. In those parts of the East, it's doing some things which look a bit like hybrid warfare as defined by Hoffman and others. There are irregular forces, there are regular forces. Irregular forces are being obfuscated, there's a bit of crime involved, there's a bit of legal warfare, the issuing of passports and all that sort of thing. Taken on its own, I would not particularly have an issue with thinking of that as a slightly different form of warfare to perhaps some of the ones that we traditionally look at. Albeit I don't think that's a new form but one we perhaps haven't seen much of recently.

The problem comes when that term is then expanded to all the other activities that Russia is doing to keep Ukraine destabilised – it's economic coercion and constant information campaigns about corruption. The desire effectively to just keep Ukraine in a slightly chaotic state such that the EU is not going to rush to embrace it anytime soon, NATO is not going to rush to embrace it anytime soon. I've referred to this somewhere else as a bit of a close and a deep battle if you like, that the close battle is what is happening in the Donbas and happened in Crimea whereas in wider Ukraine and indeed beyond Ukraine in the rest of Europe, that's more a case of a form of statecraft and that's not me legitimising it, it's a form of coercive statecraft but using modern tools, using the digital revolution to take up opportunities, whether that's cyber-attacks against the power grid in Kyiv or whether it's social media enabled information campaigns.

I think we run the risk when we start to blur these things and I'll give you a line to move you on if you wish to, and then, try to shoehorn that model into other states' activities whether that's China, North Korea, Iran, or somebody else. The problem is that we are taking what was, perhaps, a useful way of looking at a particular conflict in a particular context and now trying to apply it to everything and particularly with this use of the word 'warfare' because I do think it's almost inevitable if you start using the language of warfare and war and conflict, that the military has to be the answer to that problem. As I said earlier, this is much more about a domestic-political response, a civil contingencies type response, as it is about a military response. What the military needs to be able to do is recognise that, for example, when they're on enhanced forward presence they will be the subject of disinformation campaigns and they have been: the false allegation of rape against German military on an enhanced forward presence and the false allegations about British soldiers at various stages. That's where the military have got to be focused, not necessarily in my view, in countering the broader information campaigns which I think are the responsibility of other bits of government.

PR: I want to come back to three parts in there, about how the branding of other states as hybrid actors, the domestic responses, and firstly, the idea that hybrid as warfare. The worry for me is how we make the division between the military effort for annexation, and what we perceive it as, which is everything else, about economic coercion, the weaponisation of corruption, we lump it all in to one area and hybrid becomes warfare, which becomes the military's problem. That's one of the biggest problems that I can see.

EL: I entirely agree. I think that the military also need to look at it. We need to look at ourselves, those of us who are ex-military, as well because we've rather embraced the terminology and the language. The recent UK's Future Integrated Operating Concept, there is again nothing particularly new in that. Those were ideas of making sure that you leverage all military activity towards an operational functional output. That's not new. We were talking about that in PJHQ in the early 2000s and I'm pretty sure it was being talked about before then. But, when we listen to Chief of the Defence Staff and the Secretary of State at the paper launch, most of what they talked about wasn't the paper. They talked about the threat from Russia, the threat from China and so on and so forth but most of what they were talking about in my view, and I'm sure others will disagree, is not the responsibility of the military.

I think your point about Eastern Ukraine and Crimea is absolutely right and that's why, as I say, for me, there is a difference between – I think it's a Henry Jackson Society paper where it's called 'Russia's new conventional way of warfare' – deliberately not using the word 'hybrid' to take it out of that language problem and say, 'What you've got here is a Russia that's operating in a particular way which is fundamentally designed to stay below the threshold of a conventional Western military response. So, in Ukraine, not a Ukrainian response but bringing outsiders in, NATO, the US, the UK, whoever.' So, to stay below there, and they do that through obfuscation and doubt and hence the whole little green men thing. 'No, no, these really aren't Russian soldiers. They're local patriotic business people,' or whatever it was. That's deliberate and was designed with that single purpose in mind. So, really the key for me is, if you want to talk about something and call it hybrid warfare, it's recognising that unsurprisingly a military is going to use deception as part of its operating plan and for me, there was a little bit of a sense of, 'Well, that's all a bit unfair,' when actually, deception is a key part of all militaries' thinking or at least should be.

Going back to an observation that you made earlier which I think, similar to this, may be a rabbit hole that you don't want to go down, is that I think for the UK, with the information space, there has been a problem that outside actual conflict it has really struggled to justify having a decent information operations capability and I say that based on my time as the CO of 15 PsyOps Group and I know the current narrative is that 77 Brigade is the answer to all the world's problems but I'm afraid the sceptic in me is not entirely convinced by that. I know a lot of good people working really hard to do some good stuff and I think there are particular individuals, who I won't embarrass by naming them, who have moved the narrative on inside MoD, so it is positive in that sense but it's something we're still fundamentally not really comfortable with. Ships, tanks, and planes, that's all fine but people doing deception and information operations, that's, very good at it in wartime, if you look at the British record both in the First and Second World Wars, people come to the surface who have the ability to do this stuff but we're very uncomfortable about it in what is being conceptualised as peacetime and there's another whole podcast on the war, peace divide issue.

PR: Does the West have it within their grasp and power to respond appropriately and have credible response options to what we term 'Russian hybrid aggression'?

EL: I think it is within the power and economics is probably key but that's challenging for governments because it comes at a cost. You only need to see the continuing debate in Germany about Nord Stream 2 to realise; Germany rescues effectively Navalny and yet, having seen that happen, a deliberate attempt to kill an opposition leader with a nerve agent, having seen that happen, Germany is still saying, 'Yes, well we might do Nord Stream 2 anyway because we want cheap gas or whatever it is' and thereby hangs part of the problem. London is still – and I'll overstate the case to make the point – awash with dodgy Russian money and until such time as governments are willing to push back in those areas [there is a lack of clear and coherent messaging happening]. Ultimately, what will constrain Putin is when his financial backers start to say, 'No, Vlad. You've got to stop, mate, because you're hurting us financially.' Yes, enhanced forward presence and deterrence has a role to play, absolutely has a role to play, but it can only really be successful if there's pressure in other areas.

Also, I'd like to think that we need to think a little bit more about our protective measures, our defensive measures if you like. We don't,

I think, still have a particularly good understanding of our critical and national infrastructure, what's critical in the national infrastructure, what's critical in the critical national infrastructure, which we might have done 30, 40 years ago during the Cold War with the Key Point Programme. We certainly, I would suggest, don't really know now because digitisation has made that 10x more complicated and I think there's almost a point where people are going, 'It's just too difficult' and I think, 'No, that's not a good enough answer. We should be looking at that' but I do also think we should be looking at education. The Latvians are doing this now, educating their young people at school in national security. Now, I'm not saying we should mirror exactly what the Latvians are doing but getting our young people to critically engage with material, that's a first step in doing all of this. So, it's not just about the punishment measures. It's also about making sure we become more resilient as societies, not just infrastructure resilience, but societal resilience.

PR: How much of a mistake is it in us classifying all these other actors as only doing hybrid operations as well?

EL: I think it's interesting. Sat here as I am in Southeast Asia, a lot of the paperwork, you see here out of think-tanks in Australia and places like that, tend to use grey zone rather than hybrid. I'm intrigued as to why that is. That's a little bit of research that I want to do. Why the language is subtly different? My hope is because people here are recognising that you shouldn't just mirror what the analysis rightly or wrongly has said about Russia and apply that to China. I think the Chinese case is an interesting one, the use of coastguards, maritime militias. Again, I think this is about staying below thresholds and I think that's where the similarity almost begins and ends. I would suggest that, in part, this may be because the PLA and the PLA(N) doesn't feel it's quite ready yet for that force-on-force confrontation and these so-called hybrid approaches or this desire to stay below the threshold will last as long as they feel that they're not capable of direct competition, particularly with the US Navy, but also increasingly Australians and others in the region.

So, this might not be a way of warfare but rather an approach to be followed to achieve political objectives in a meaningful way in the short term until such time as China becomes a meaningful military power once again but there is a real problem with China with all these things. The IRGC today, I think it looks more like Hoffman [would have envisaged in his original 'hybrid' warfare paper] than pretty much anybody else does, certainly more than Russia does in terms of combining conventional, irregular terrorism with a criminality. That's kind of the way Iran seems to be operating. North Korea's ransomware attacks and indeed its general use of cyber seems to me to be much more about a state that's desperate to get currency because it's so completely and utterly isolated and that's one way to do it. That's why the attack on the Bank of Bangladesh a few years ago has been linked to North Korea. That was absolutely about getting cash. Is that about warfare? I'm not so sure.

Reflections

Warfare below the level of conventional combat has been a constant part of interstate competition for millennia. It has been known by a variety of terms throughout history but – like many aspects of warfare – can easily slip the mind of those less familiar with any history of warfare. Even when major conventional campaigns are taking place, irregular warfare, competition leveraging some aspects of military force, non-lethal coercion activity, and special operations continue to take place: probably as part of a major campaign but also with and by other actors seeking to exploit opportunities forded by the distraction of other belligerents. Whilst the intensity of such activities will vary depending on location, actors, ambition, resources, and time, it would be a brave person to discount such actions from having constant residence in the international system. As noted in the above discussion, successfully overcoming such activities requires an ability to recognise when they are being used against you, reacting appropriately (for example, not over reacting, but not ignoring what are important aspects is signalling between states), and resourcing those reactive measures with both realistic ambition and considered strategy, as well as empowering the right people to utilise the correct levers of power to ensure that the ambition can be delivered.

Most states employ hybrid techniques to achieve influence and support their national interests. The Wests' favourite tools sit within the rubric of a 'Rules-Based Order' as invented and perceived by Western states. But to the rest of the world (the other 85 per cent by population), these levers conform to various definitions of hybrid warfare. We need to be wary of our own imposition. Of double standards if we (the West) are not to continue to alienate the rest of the world.

Hybrid warfare is but one of the 'new' definitions of warfare that were raised between 2003 and 2022. If Frank Hoffman started the trend with Hybrid in 2007, it was followed by a wave of theories from Fourth Generation and Irregular warfare (Rid, 2009), Proxy warfare (Hughes, 2012), Non-Linear (Galeotti, 2016), Grey Zone (Echavaria, 2016), Surrogate warfare (Krieg, 2019), and Shadow warfare (McFate, 2019). None were rejected by militaries or political leaders as lacking utility or fallacies. All had a fair amount of hype associated with them. Yet few have survived contact with the national security community after the renewed Russian invasion of Ukraine when conventional war was deemed by mainstream societies to be back in vogue again.

An argument could be made that this is simply militaries looking for the next big thing in order to stay ahead, but it is also a useful example of how Western militaries seem unable to keep two competing ideas in mind. To most Western military personnel, warfare is either conventional or it is being waged unconventionally. The discussion becomes polarised between two perceived extremes. One might ask, why is it not possible for adversaries to employ both conventional warfare and unconventional levers together? Indeed, the real question is why would they not? The idea of non-conventional challenges appearing on a battlefield, or an area of tension, is neither new, nor predictable. But it is reality. And like conventional tactics, unconventional techniques will change depending on context, threat, and a myriad of other factors – just as conventional ways of war might, in some circumstances, favour a recce-strike approach in one moment, and combined arms, armour heavy fighting the next. Unless Western national security professionals prepare themselves for a range of actions and activities by adversaries, they will be defeated before the first round is fired.

7

The Media and War

Whether the media actually shifts and changes the culture of its audience or is simply a factor in society's evolution is a matter of dispute. Yet the media has a long and august history relationship in reporting on wars and combat. The view of the media by the military waxes and wanes between despisal and adoration but between wars where the nation state is involved, in periods of peace, some reporters have an exposure to war that militaries lack. The access of some media personnel to conflict outside of a Western sphere allows them exposure to wars as fought by others; opinions and observations that can be absent in Western military discussions. These reporters and war correspondents bring perspectives that are necessary, albeit sometimes difficult for Western militaries to digest. Thus, observations from the media on the Western way of war is instructive and, indeed, essential to develop a rounded view on the topic.

Informing the public, whether the domestic audience, potential allies and suppliers, and adversaries, has always had a role in war. Its relative importance is contested by evangelists and sceptics alike: it depends largely on the way leaders seek and are able to exploit information to further their own ends. It could be argued that mainstream media has developed a more prominent role in wars since the start of the 20th Century. French professor, Arnaud Mercier, notes that evolving four factors have made warfare a media spectacle: Photography and stage-management; Live streaming technologies driving the need for newsworthy stories that maintain personal profile and raise issues of perspective; media pressure and media globalisation, altering the way in which political and military actors undertake propaganda and information operations; and the changing views on state censorship, a necessary requirements for operational security but regarded as ethically wrong in the West, requiring states to find new ways of controlling journalistic access, and the media more generally.[1]

There is also a view that since given the spread of media outlets available on wars and combat, the mainstream media have sought ways to differentiate their stories from simply reporting events. As access to battlefields has increased, and independent, individual views can be aired from participants and civilian parties in real time, the media has had to move away from simply reporting activities as they occur and official government announcements. Whilst having its roots in the classical world, mass persuasion was honed and exploited very successfully in both the UK and the USA during the Second World War, when propaganda was used to garner and maintain support for the war effort. These activities were government led, relying on information gathered and provided by the state. Indeed, the funding for these enterprises was prioritised by governments.

The Vietnam War demonstrated to political and military leaders the dangers of unrestricted reporting from war zones on public opinion, something which the UK government learned a good deal and shifted their own approach to war reporting that was demonstrated during the Falklands War in 1982. The ability to keep reporters away from the battlefields, and censor their output, allowed the UK government to control the narrative. Further change, largely derived from technology (camcorders, sat phones, commercial access to war zones and so on), allowed reporters increasingly unregulated access to war zones: governments understood that simply keeping reporters out was no longer feasible. By 2000, press pools and media personnel embedded with military units became normalised practice.[2] So too did the emergence of a 'peace-orientated' media: an opportunity for the media to set their own agendas and shape both public opinions on wars and the cultural relationship between society and organised violence.[3]

Thus, the interaction of the mainstream media and forces operating in wars is complex and fraught with tension. Historically, different national

---

[1]   Arnaud Mercier, 'War and the Media: Consistency and Convulsion', *International Review of the Red Cross*, vol. 87, no. 860 (December 2005), 629-659. https://www.corteidh.or.cr/tablas/a21918.pdf.

[2]   M. Taylor and M. Kent, "Media in transition in Bosnia: From propagandistic past to uncertain future", *International Communications Gazette*, vol. 62, no. 5 (2000), 355-378.

[3]   Vladimir Bratic, 'Examining peace-oriented media in areas of violent conflict', *International Communications Gazette*, vol. 60, no. 6 (2008), 487-503.

militaries have had very different relationships with the media, depending on the context and experiences of both parties, as well as the national strategic culture that tends to be accentuated and amplified in the media / military relationship.

Western media agencies have tended to be circumspect about the military and their operations reflecting, perhaps, a different approach to the release of information, the tensions over the need for secrecy versus disclosure, and an inbreed distrust between the two professions based on a long history of uneasy experiences. Caricaturing these parties sets them poles apart: the conservative military with a culture that reveres its own history, heroism, personal sacrifice and the glory of battle opposite a liberal but caustic media world devoted to sensationalism, modernity, and a doctrine of 'truth (airtime) no matter the costs'. Whilst there are elements of truth in this parody, it is also a broad sweeping generalisation that has no place in a debate about the relationship or their professions.

Both parties believe deeply that they have been doing the right thing for the right reasons, but each fails to accept the other at face value. For militaries, the media are determined to find an exposé that paints military personnel, their leaders and actions in the worst light possible whilst military personnel see themselves doing what they have been required to do by their political and military leaders in particularly harrowing circumstances. For reporters, the military are often unhelpful, obstructive, tools of the state that are designed to deliver death and destruction, and – in the eyes of the media – lack compassion, a moral compass, and possess a ruthless streak that neither party recognises in themselves.

Within the UK military, tension with the media is extremely high. Poor experiences since 1914 have led to a cultural allergy to dealing with reporters. Distrust has been the default setting for both media and military personnel. This is not true of many militaries and their national medias, but it is worth examining as a case study of the extremes to which a media-military relationship can descend. The following conversation took place with veteran reporter Lucy Fisher, an author and one-time defence editor of The Times of London. Lucy's experiences in dealing with the military, particularly at a very senior level, were worth recording and recounting if we are to maintain a broad an encompassing understanding of the Western way of war.

What Does the Western Way of War Mean to You?

Lucy Fisher: To me, the distinctive thing about the Western way of warfare is the values we place at the centre of the way we do things – everything from rules of engagement on the battlefield in a kinetic scenario to our refusal to enter into the sort of dirty tricks that we increasingly see played by the likes of Russia and other hostile nations. I think that is the guiding principle of the way that we conduct warfare, and something certainly in the UK I think that there is consensus on, and I don't think we are likely to see any change to in coming years.

Peter Roberts: It is interesting that you raise the values aspect; fascinating in fact because if we look back through the history of British warfare, we haven't exactly abided by those values in many times, have we. Indeed, the British way of warfare has often been that we have been the rule breakers. We've been the ones that have used the very edges of the envelope in terms of legal, moral and ethical behaviour. It's this sort of reimagining of Western values since the end of the Cold War perhaps that has made us think this way. But it's now deeply embedded, as you say, in almost the culture, the feeling, the perception of how we would use force, right?

LF: I think that's absolutely right, and you are right to point out the fact that it hasn't always been that way. But I think it is a constraining feature of certainly the political discourse around how we can use our armed forces now. I think it is increasingly a problem when we face a barrage of information operations, how we can both counter those, but also questions about how we might engage in that sort of sphere of warfare or tradecraft ourselves. Increasingly as we get more and more cutting-edge, sophisticated technologies that have a covert function, I think there is growing concern about how much accountability there may be for the services, and how much the government and ministers will be able to tell the public about increasingly covert capabilities, and how much tolerance the public would have for the use of those.

PR: That's great, we've twisted the discussion towards the politics [of war and the media] quite early, or the personalities (in many ways) of the politics, because you get the feeling as a military officer that the military is almost the prime minister's tool. Prime ministers come in wanting to end wars and withdraw from things, and then suddenly they want to define themselves so they get engaged in a war somewhere and start using the military [for their own agendas]. Their view of the utility of the military instrument tends to change quite a lot while they're in office. You see this on both sides of the House [of Parliament], and you see it with Tony Blair and Gordon Brown, you saw it through David Cameron, Theresa May and up to today. The military becomes one of those tools that prime ministers seem to feel defines them, how they use it, what they engage in. This is a military officer's perception, but do you think there's something to that?

LF: I certainly do think there is something to that. I think for me there's a really important point in recent history we need to mark that has changed how the use of the forces has occurred in a political sense. That for me was that key vote in 2013 that David Cameron lost, when he put to the House of Commons the idea of greater military intervention in Syria. I think there is a sense now that after the expeditionary conflicts in Iraq and Afghanistan – which frankly I think most MPs would rather not talk about, especially those that voted for action – I think there is a sense that prime ministers must feel constrained to seek the approval of Parliament to go forward with military action. I think that is a change. They're not as free to use it as their tool and as part of the arsenal they have, being in Number 10 [Downing Street – the residence of the British Prime Minister]. They need that consensus to have Parliament on board.

PR: It's interesting, isn't it, because the loudest voices about defence in Parliament tend to be almost not the ministers, but those on the defence committee, or the 54 or 56 Members of Parliament who have previously served in the military. That is the highest number that I can remember for a while. Do you get the feeling that the discourse in Parliament about the military and defence is improving? Is it better than it was previously? Or do you think most of these people have experience at the lower levels of command in the military – do you think it's denigrated down to the tactical engagements rather than thinking about grand strategy and Britain's place?

LF: Well, I think it's less in depth than it used to be. As you say, there is a high number – I've got the figure here, there's 7 per cent of MPs who are currently or previously regulars or reservists in the armed forces. That is higher than the working age population, which 5 per cent are veterans. But that's it, as you say. They tend to be people who may be reached the rank of captain before leaving the forces. It certainly seems to me that it [being a veteran] is not something that is a vote winner. So that's what most MPs are thinking of: is it something that directly affects their constituents? If they perhaps have defence industry factories or manufacturers in their constituency – that is a reason for them to play up the issue. Others have more in-depth knowledge, as you say, of foreign policy, the strategic context and may have a personal interest that may incentivise them to speak up. But overall I think we have to put this in the context, never clearer in [this] a year of a pandemic where health is front and centre, where education is also vying for air time, and where welfare is top of many people's priorities. I think in terms of what drives politicians is what matters to voters, and secondly that in turn drives where the funding goes. The Defence Select Committee of back benchers is a very strong force for trying to hold the government to account, not least being led [at the time of the interview] by Tobias Ellwood, former captain in the Royal Green Jackets. He's also a former minister, which means he knows lots of the tricks, he knows some of the skeletons in the closet of the MoD, so that makes him a very useful person to be able to speak freely about what's going right and what's going wrong. But I think by and large it's not the calibre of discussion and it's not the depth of knowledge that we had in certainly the decades after the Second World War, and perhaps in the midst of the Cold War. You had many people who'd got to more senior ranks of the military.

PR: Yes, and who then went on to occupy political positions. It's always interesting to reflect on the level of discussion. Those who serve in the military are always frustrated that the military is never top of the agenda. Usually, it's well below health, it's well below education, it's well below welfare. It certainly sits down the [political] priorities list, but then it's something that is one of the few things that is identifiably British. Apart from the NHS (National Health Service), there are the Royal Family and the armed forces, but there aren't many other things that are sort of identifiably British these days, certainly for the United Kingdom. It always strikes me as strange that politicians will hark on about the utility of the military and what great fans they are of it, and yet they fail to be swayed by arguments of the level of threats, what the military are facing, how they need to be updated – I mean it just doesn't wash against voting for a new school versus some new tanks. Would you agree?

LF: I think that's right, and I think that primarily for two reasons. The first is politicians are highly risk averse when it comes to any operations or missions that risk British lives. I think you're right that the armed forces are celebrated and revered in many ways, but I think there is a huge sense of politicians feeling wary and cautious of sending the military into dangerous missions and I think that follows on the [Royal] Wootton Bassett effect of seeing those flag-draped coffins of young men and women in the sort of height of the Iraq and Afghan wars. There's very little appetite to actually use our military to do anything particularly dangerous these days. I think the second thing is the changing way that warfare is occurring. We are not seeing the same sort of hard power threats perhaps as we used to, or the way in which hostile nations are exhibiting aggression is different, it's grey zone. I think that's something which the public is ill informed about. It's difficult to explain to people how some of those tactics are working, and the attribution is also difficult, whether it's information operations in the use of bots on social media, whether it's cyber-attacks where it can be very difficult even though there may be a motive to sort of pin down exactly the nation-state behind something – indeed because they may be using in-house kind of or outsourced sort of criminal gangs to achieve the effects they want. You know, the increasing use of proxies. I'm interested that we're beginning to hear more from ministers like Ben Wallace, the defence secretary [2019-2023], talking on record about the Wagner Group, but again it's difficult if it's not a nation-state's army doing something on the ground abroad, but a private company, to explain to the British public why we need to invest in defence when it seems so sort of disparate and diffuse the way that other hostile states may be acting. I think that's another reason many politicians demur from getting into the subject at all.

PR: It is quite strange that there seems to be a narrative and understanding across the political class that everything only happens in the grey zone, and it's really strange when you contrast that with the lived experience of conflict, from Ukraine and Syria and Georgia, and even in Libya where actually we're seeing quite significant kinetic action. The experience of the Israelis in Syria is not that cyber and information is doing amazing things. It's failing to live up to expectations in many ways and having to revert to hard power – I just have a feeling that the political class is relishing the conversation about grey zone operations because actually there's no real risk in employing UK forces in the grey zone. The problem would be if the politicians acknowledge that fighting is still kinetic to a significant extent and results in hundreds of dead people, thousands of dead people all around the world every single day, and for a politician to admit that and in fact they're going to need to drop bombs and kill people is again not something that's going to win them votes. It seems to be easier for them to talk about grey zone and that being the new future, even if there's no evidence to back that up.

LF: I think you make a really good point. I think that's sort of nails it. It is easier for politicians to talk about that, and I also think there is so little appetite following Iraq and Afghanistan to send British troops into expeditionary conflicts. People don't see why we should be in Syria or politicians don't want to put their head above the parapets, or not many of them – to be fair there are some who have made the argument for doing this, but they haven't persuasive or the majority, let's say. There isn't the sense that we want to get involved in conflicts like Syria, like Libya, so I think that politicians don't feel any need to be drawn into a debate that probably doesn't hold any benefits for them.

PR: What do you think would change that for them? Do you think that a conversation about the Baltics or Scandinavia, about Russian presence and special forces walking around Sweden or Finland, or operating in the North Sea or the Irish Sea – what would change this view? Because it seems a remarkably distant conversation. They can talk about the grey zone and do more in cyber and let's informationise this and data and invest in AI and perhaps some satellites, but I mean, really, it stops short of any difficult decision. What is it, do you think there even is something that would be a catalyst for a change again, a bit like 2013 was for the vote?

LF: That's a really good question. I certainly think if we saw more activity in Europe that could be a catalyst. I think, yes, there's long been a question about whether the US and in particular the Trump administration would consider some form of kinetic Russian activity in Estonia, something they would feel beholden to be involved in – obviously their obligations to NATO very much up in the air on that one perhaps. But I do think that in Britain the government would not stand by if we saw something. But again it comes back to perhaps the way we've seen Russia acting, using proxies, choosing tactics and choosing missions that are harder to attribute, that don't allow the nations of the West to stand up together in unity and condemn them or indeed take further action. I think it's less likely perhaps that you'd see an adversary choosing that course because to me it seems still a strong alliance in the West that would react very badly to that.

PR: But if – we're all talking worst-case scenarios – a Trump second term that potentially took the US outside NATO and left them with their bilateral arrangements in the Baltics and Scandinavia, with the Northern Group and the UK and perhaps bilaterally with a few others, that would change the landscape fantastically – fantastically not used in a good way – but it would change it starkly across Europe in terms of the defence conversation, right?

LF: Absolutely! It would just tear up every presumption on which our sort of defence posture is based. At the moment there's a two-pronged strategy that the UK and other Western nations are sort of going along, which is firstly to try to preserve NATO and preserve the appearance of it as unified, as far as possible, even though there are huge degrees of nerves about President Trump's ambivalence and worse still the many reports of how close he's come to directing officials to try to withdraw the US and as you say a lot of concerns about if he wins a second term whether he would actually go ahead and do that. I think at the moment, for the past few years, we've seen the UK and other nations just try to keep NATO together as far as possible, to deny in public any sense of fragmentation, while at the same time looking to deepen and build new alliance, clubs within the club, and indeed clubs outside the club. Certainly you've seen within the UK moves to strengthen the Joint Expeditionary Force with Scandinavian nations and the Baltic states. Also, looking further afield to the Indo-Pacific, deepening the defence relationship with Japan is particularly striking. Also I hear much more in political circles the 'Five Powers' mentioned – UK, Australia, New Zealand, Singapore and Malaysia [the Five Powers Defence Agreement]. I think with the growing realisation of an increasingly assertive China posing ever more challenges for the UK and the West, certainly the UK is looking further east as well. But whether those new alliance or sorts of mini coalitions are any match for NATO, there's no pretence there. Nobody thinks they're anywhere near the strength and interoperability on a military level that NATO has built up.

PR: But China is on the horizon – no one is doubting that. The level of influence might be falling slightly in the UK but it [China] is financially intertwined in most of Europe. China [state companies] own several critical ports, it owns key utilities, it runs critical national infrastructure in many European capitals let alone across the countries. It's a rather fantastic problem to get into and the UK just seems to have turned a corner in terms of its relationship with China. It might be Hong Kong and the national security law, it might be a huge number of things, but it feels like the UK political discourse is somewhere ahead of Europe in terms of that. But I haven't heard much from the other side of the House in the UK about it. Do you get the feeling that the Labour Party feels a similar nervousness about China's influence in Europe?

LF: Again I think you have to go back to politicians' first prerogative, which is winning [and maintaining] power. For the Opposition I think there is certainly a sense that they don't want to be painted into a corner too soon. The next election is likely not until 2024. It's a long time away to take policy positions, and I think there are certainly concerns about China in some quarters of the Labour Party. Whether they want to come out now with firm policy positions, the best of which they worry could be nicked by the Conservatives and any sort of policy pronouncement missteps could be held against them at the next election means that I'm not convinced we're likely to see them standing up on this issue with great vigour in the short term. it does seem to me that it's an issue that has been most zealously seized upon by the Tory right – Iain Duncan Smith, being the British co-chairman of this interparliamentary alliance on China. I think it will be interesting to see how the China debate moves forward, whether it becomes an issue of the right or whether Labour gets on board with it. I think there are concerns on the Labour front bench about how the UK has allowed itself in lots of strategic areas of the economy to become reliant on China, and supply lines in strategic goods where we [the UK] are a net importer, and more than 50 per cent of our imports come from China. I think they [UK parliamentarians] are beginning to ask questions around specific issues that leave us perhaps vulnerable in diplomatic terms. But there's still some way to go, it seems to me, before they catch up with the most concerned of the Conservatives.

PR: Now I want to move across to the second big topic that listeners wanted me to gather your views, and this is the media's relationship with the military. When I was in the military, the one thing we wanted was lots of good news stories about defence everywhere. It was what we and senior commanders wanted and we could never figure out why they were never there. There were good news stories all over the place – rescuing people and seizing drugs and propping up dams and driving ambulances and oil tankers. But they never made it into the media: it was always frustrating. There's a huge history that would be explained to you right back to the shelling of Scarborough in the First World War by the German fleet. There was the story about the bombs being dropped in 1982 and the BBC effectively telling the Argentinian forces that they had their fuses incorrectly set. As a result, this relationship with the media, as a military person, it felt hostile. Even before I came into contact with any journalist, it felt as if I had to protect myself, protect my service, that they were out to get you. It was almost within the culture of an organization, a military organization, that this relationship existed. It struck me as deeply unhealthy, and it wasn't really until I left that I started to understand that relationship between the media and the military was poor largely because of the military's behaviour, rather than anything else. Now, that was the way I would characterise it. Perhaps you would probably say something completely different?

LF: Well I think what you say is interesting, and I do recognize the sort of wariness you speak of. Many officers don't want to see junior colleagues in any way shown up or bringing the services into disrepute by perhaps misspeaking or characterising something in a way they felt was infelicitous. I think that actually it's – what we're talking about now, the relationship between the media and the military – not entirely separate from the relationship between the military and the politicians, because for me that's where I'd often see the services would like to do more but at the same time they have to agree that with political overmasters who also have their own agenda – and the great example is right now we're in the midst of this integrated review into foreign policy, defence and security, in which each of the services would very much, I think, roll the pitch for what they'd like, their own recommendations, ambitions and so forth. But from a political standpoint, you've got ministers who don't necessarily want the public to see services fighting between them so they want to coordinate and present in their own terms what they think that the military as a whole should have. I can understand why people feel nervous dealing with the media but I also think it's not necessarily at the core of military thinking that they are accountable to the nation because we are a democratic nation and if they are sent to war in the United Kingdom's name, the public have to be brought along and have to understand what they're doing and how they're doing it, also that they are accountable because they are funded by taxpayers. I think that is a link that is clearer in other parts of the state that is not always at the forefront of the services' thinking. But I'm interested, Peter, in your time do you think that the relationship between the military and the media now is better or worse than it has been in recent decades?

PR: I think it's better, but I think it's better primarily because of the way that the media drove that engagement in Iraq and Afghanistan. I genuinely think that had it been left to the Royal Navy, who are terrible at media relations, and the Royal Air Force, who aren't much better, the British Army did want to show the reality of combat on the ground and what they were getting into, and there were brilliant, embedded journalists who deployed with them. For the troops on the ground, right at the front line, at the cutting edge, I don't think any of them are scared of showing their views, of showing what it's really like. They are proud of what they do, and rightly so. They perform these fantastic, amazing feats, but somehow it [the willingness and openness to engage with the media] gets quashed out of them at higher levels, the nervousness over what a bad news story is, because in many ways I think the military see the relationship with the media, or the media's job, is to report the stories they want to report, and not those that senior officers want reported. That's just not the role of the media! It's frightening that they've got into this view that the media should be reporting what they want them to do, rather than either the reality or just stuff that is not good, because that returns to what you said originally: one of those values is that we are open and we have freedom of expression. We don't live in Hong Kong; we're allowed to say things. So we shouldn't regard people who start to bring up complaints as whistle blowers to be shielded from the media. We should be proud and open about what we do. I do think we're in a slightly different place. The military are and I don't think will ever change this relationship they have with the media, which I would say is adversarial. I don't think there's any other way to talk about it. In the UK it's adversarial. The military twitter-arti are terrible about condemning anyone who doesn't follow the lines that they want, of looking through rose-tinted spectacles at every bit of capability they [the British military] got. I do think the relationship is better than it has been, but I don't hold out much hope that we're going to see a huge change to the way that the military get frustrated at the media. There is a helpful side and a hurtful side to it: the military in many ways feel that they have a right to use the media. I don't know if you feel that this is something you've seen, but if you look at the latest defence review and the leaks of material they've been given (no more tanks, cutting frigates, cutting aircraft whatever it is), these are all attempts by the single services to spin their own place, right, to scaremonger to a certain extent.

LF: I think you've always got to be aware as a journalist that people are usually telling you things for a reason. There is usually an agenda there and you have to balance that. At the end of the day you have to go with the information you are able to gather, and I would certainly say that in my experience as a defence journalist I have found it a far more opaque world than politics, and I think that's for lots of reasons, including – I'm making here a very sweeping generalization – the characters of people who perhaps enter politics are very high risk takers, and people who very much value information exchange and understand that from you, the journalist, they can also bat ideas and try and get a sense of their other rivals might be up to and an appraisal of the opposition's strategy and so forth. I think there's just a lot of more sort of suspicion of what the media is up to in the military. But I do think that Twitter is not the real world. I certainly don't let the sort of 'buffs and toughs' of Twitter bother me and in fact I just don't waste my time engaging with them. I think more widely off Twitter there are lots of people within the sort of military and wider defence community who do see the value of keeping defence high on the agenda of the media because they need to make the argument about why it should be better funded. We've seen the 2010 and 2015 strategic defence and security reviews levy pretty severe cuts. We've got equipment plans now that have multi-billion funding black holes in them. Thus, unless you're willing to sort of make the argument about what you're doing, what's going right, what's going wrong, you fall down the politicians' agenda and the public's agenda and that's when you begin to get squeezed in a financial sense. I'm always sort of more than happy to try to explain to people why it's in their interest to have engagement with the press. I think that you mentioned some particularly difficult examples, such as the Falklands, the case in which the BBC World Service was accused of in effect tipping off the Argentines about issues with their bombs which the defence attaché in London heard and therefore was able to correct. I don't think any journalist wants to endanger national security or personnel lives in any sense, so it's definitely something I feel very strongly about myself and all the other members of the Defence Correspondents Association, who I know pretty well. I think beyond that there are a lot of wins that the media have achieved for the military, whether it's looking at something like really shining a spotlight on the problem with 'Snatch' Land Rovers in the Middle East and how they have become these mobile coffins leading to the deaths of so many service personnel. Keeping the focus on that issue led to their eventual replacement, perhaps not quickly enough, but definitely played a part in making sure the politicians back home were aware of that issue. I think also over the weekend we're seeing the Defence Secretary and the Home Secretary talk about bringing another 100 Afghan interpreters to the UK. I think that's a great example of media-military cooperation. It's been some fantastically committed people like Colonel Simon Diggins, a former British Army officer who served alongside some of these Afghan interpreters, who refused to let the matter lie. His concerns about these men living under severe threat under the Taliban in Afghanistan – he really felt and many of his colleagues felt, the UK owed them a duty of protection following their service, partnered, I'd say especially with the Daily Mail, which has campaigned non-stop on this issue, they've forced politicians to change the rules and allow more of them to come over. So the media is not always helpful and it's not there to be helpful, but it can effect changes that are of benefit to the military.

Reflections

It is worth noting that this interview took place well before the Russian invasion of Ukraine in 2022, but after the initial illegal invasion in 2014. The early part of the conversation thus reflects some of those feelings within defence circles about the future of war being characterised by Grey Zone or Hybrid warfare (a topic covered in chapter 6). It is interesting to reflect on the 'unthinkable' scenario of further Russian military action in mainland Europe; something regarded by Lucy Fisher (a political and defence expert in her field) as a potential trigger to momentous shifts in attitudes to defence spending and debate. What is clear after February 2022 is that politically, in the UK and elsewhere across NATO member states, there was considerable consternation and announcements of change. Yet, some two years after that occurred, the defence community has seen very little additional investment or capability growth. Indeed, most NATO states have seen further cuts to the conventional military capabilities despite increases to their funding (on paper at least). With the notable exception of Poland, the Baltic states, Sweden and Finland (as new members of the alliance), NATO remains a shadow of its former self. Having started this chapter wondering whether mainstream media changed the culture of a state (or simply reflected common views), particularly in reference to national security, there seems little evidence that mainstream media has exercised either role since 2022. Of course, this might simply be because that defence correspondents are unable to fight their own battles with editors for commentaries and type-space that advances and educates the wider public in a better understanding of defence and security.

The discussion above was also interesting because of the seeming ambivalence of mainstream media towards social media, certainly in matters of national security. There has been a good deal of excitement and hyperbole about the role of social media in conflicts between 2022 and 2024.

Whether in Syria, Ukraine, Yemen, Iran or Ukraine, the discussions over the relative importance of social media have almost been more widespread than the use of it as a means of communication, propaganda and influence. The amount of national security coverage and commentary within traditional media has also increased but with little impact on a 'national debate'. The column inches given over to war reporting and analysis seems to have increased markedly, yet the impact on government policy has been negligible.

Perhaps our understanding between the relationships between the media, information, conflicts and the public is no more mature and complete than it was in 2001.

8

Military Procurement

The theory to buying military equipment is simple. The people on the front line decide what they need to succeed and buy it from industry who deliver the right equipment in the right place, and the right time, in the right quantities and with excellent quality assurance, and at a price that all are content with. One might expect additional urgency and rigour during the process: military equipment is not something to make life easier, or for convenience; it is often the difference between life and death. That is not hyperbole. From personal weapons, body armour, and communication equipment to fighter aircraft, jets, tanks, warships and submarines, these facets of military life are designed for an environment where it is, quite literally, "kill or be killed". In those circumstances people want the very best they can have and sometimes the public and political leaders of that state will feel the same way. But only sometimes.

During periods of major conflict, usually associated with existential threats to the state, nations become adept at turning their entire focus, economies, and manufacturing base towards wartime production and survival. In these periods it is rare for the bureaucracy of acquisition to disappear completely, but it is tinged with more realism and urgency. There is less focus on long term technological research that could bear fruit in a decade or two, and more alignment over priorities. Industry and contractors are still required, by their owners and shareholders, to make profits – sometimes a greater percentage than in peacetime – but delivery can be easier as the requirements do not tend to change markedly as priority is placed on speed of delivery rather than exactly matching the capability and price parameters.

These three aspects lie at the heart of military procurement and acquisition: the balance and priority being placed between capability, price and speed of delivery. Purchasing any piece of kit (whether military or not) requires a balancing of these three facets of each contract. 'Which is the priority?' has different answers depending on the strategic circumstances each nation finds itself in at any given moment. This does not appear to be an overly complex conundrum, however.

Yet it is hard to find a single military procurement programme anywhere around the world that delivers what is promised, on budget and on time to the satisfaction of those using it. Sometimes that is because a military cannot afford what it really wants and must settle for second (or third) best. But most of the time it is simply because of the mismatch between requirements, expectations, industrial base, timelines, cash, political expediency and decision-paralysis. These frictions are amplified during periods of 'peace', meaning that the corporate memory capable of reverting to a more effective (although not necessarily efficient) system capable of equipping and supplying a military force is lost. The longer the period between periods of wartime industrialisation, the worse the malaise and failures of procurement systems become.

As a result, no one seems to do military procurement and acquisition well. In open and honest conversations about buying kit for militaries, there is a sense of huge frustration from personnel involved in the process from around the world. On the military side, the soldiers, sailors and aviators on the front lines (and their engineering support staff) can't quite believe that all the cash invested in equipment for them delivers so little when so much of it is brittle, fragile, ineffective and rarely fit for task. Their leaders seem to battle against bad news stories about what is being delivered while signing contracts for even more expensive kit in the future that will only come be fielded decades after they have retired. For industry, the frustration comes in dealing with a constantly changing set of managers and leaders in military and government who have no understanding, and little interest, in what it takes to deliver a proper partnership that might ease the tensions, while dealing with shareholders who want a higher return on investment than politicians and their militaries are willing to provide. As a result, these tensions result in a relationship fraught with misunderstanding, bad behaviours and tattered reputations.

This complex conundrum is not well supported by the people within it. It is hard to find anyone not trying to make the systems work best across actors in the system, to get the best kit to the military personnel on the front lines. But they all seem to fail – with increasingly disastrous results over periods of long peace and tighter financial budgets. There are two issues that mean that this collective works against itself: the first is how different parties use different (and sometimes increasingly divergent) languages resulting in an inability for each actor to understand the others, inevitably leading to increasing divergence and incoherence; and second the cultures of the organisations and actors. This latter issue was addressed by Professor John Louth in a conversation we had over procurement and acquisition behaviours.

John is an author, adviser, business executive and researcher, working in the spaces where governance, politics, the military, commerce, the third sector and the public converge. His passion is in trying to understand and explain the complexities of defence capability generation and management, and that comes from a multifaceted public-private enterprise that embraces prime contractors, technologies, SMEs, the armed forces, government decision makers and families. John was director of the Defence Industries and Society programme at RUSI, a specialist adviser to the UK House of Commons Defence Select Committee, and was the UK representative on a pan-European research body for the European Union on defence capability and readiness. Prior to all of this, in a former life, he served as an officer in the Royal Air Force for 16 years. His latest book on defence was published in 2019.

What Does the Western Way of War Mean to You?

John Louth: I think in responding to that question we get an early glimpse perhaps at some of the drivers for understanding defence acquisition as well. In responding to what is the Western way of warfare, I think any smart person would say, well, it depends, and it depends on the perspective that someone is taking. If we take a traditional, sort of post-'45, Second World War perspective, the Western way of warfare is through large treaty organizations, NATO being the principle one of course, in direct adversarial competition with Russia or the USSR, rather, and the Warsaw Pact. It was, if you like, prosecuted through the Cold War with pre-positioned forces, significant forward stockholdings of spares and, in many ways, it could be characterised by militaries letting industry, whether it's public ownership or private ownership, know what they want. Defence industries [across NATO states] built stuff, the militaries operated stuff, and had spares to enable that stuff to keep going during a potential war in Europe – a timeline of two, three, or four weeks perhaps before it all went really, really terrible or one side gave up. In terms of a procurement system within that traditional way of Western warfare, hugely static, bureaucratic and very much a model of military decides, industry builds, and that's the end of the relationship in many ways.

A second response within the 'it-depends' rubric could be one of morphing on from that sort of adversarial block-on-block stance. We then had the out-of-area challenges, perhaps offered by Iraq and Afghanistan and, to a lesser extent, Libya, where there was a need of continuous supply and significant involvement with manufacturers on the front line, in terms of maintenance and modification, with a lot of needs being satisfied through urgent operational requirements that, in essence, bypassed the static, bureaucratic procurement systems of the West.

A third 'it-depends' could be thinking about that what we need today and near-term tomorrow is still likely to be technology-rich capabilities, because it was always the Western way of warfare that the West would use niche technologies to overmatch Russian mass forces, if you like. So even today the capabilities are technology-rich, but they feed much smaller forces, and of course go across many more domains, which we are only just starting to understand – cyberspace, broader notions of information and influence and so on. That of course is a significant challenge to both the conventional force generation model through, if you like, static, bureaucratic procurement and the changes into UORs (Urgent Operational Requirements), so the Western way of warfare could be any of those. Take your pick! But the Western way of warfare tomorrow is probably going to be more towards the third.

Peter Roberts: What you present there is three very different concepts of warfare but it's also different concepts of the relationship with industry and the relationship with technology. I mean, in the first one you talk about, for the vast majority of the Cold War up until the early 1980s, the West wasn't highly dependent on a technological edge. It was in many ways playing, or it felt it was playing at a lower technological level than the adversaries. The Warsaw Pact was deemed to have better kit, it was deemed to have more of it, and it was only when AirLand Battle doctrine came in, in 1982, that this switch was made to use technology overmatch, which was something you then see coming back in the third form of warfare you described. The second one, I think, I can't figure out if it's an anomaly or is actually a pattern of changing relationship with industry. Does that sit at the heart of this dynamic? A relationship with industry where, as you mention in the first scenario, military decides [what it wants], and industry delivers. That second scenario is about the military crying out for something and industry giving them what they can that's immediately available. The third scenario, which feels far more like a meaningful partnership, a relationship that industry starts to move into the centre of the military lever of power.

JL: Yes, I agree with much of that. I think where I would characterise it, perhaps, is the traditional NATO-Warsaw Pact model really did play to the notion of military exceptionalism. Everybody on a base like RAF Brüggen in Germany, was pretty much in military uniform from the chef to the dentist to the doctor. In fact, the only people who weren't in uniforms perhaps were the cleaners, the carpenters and the teachers. There was an entire community of men and a few women, but mostly men, in unform who'd be fighting the war in the short term before they were over-run or common sense came back to the fore. That really put industry in the stance of being mere suppliers and providers. Once everything was in the maintenance unit at an RAF Brüggen or forward stored with the force squadrons, that was there for the military to use, we can forget about our industrial colleagues and friends, and in fact they're probably going to be destroyed anyway as our chums from the Warsaw Pact come west. That is a completely different model from the need for industry modification and front-line maintenance with our out-of-area challenges and you can see how the number of people in overalls rather than uniforms grew in relation to their military colleagues once we get into Iraq and Afghanistan. In a way that was very much a hinge between a model we had before (that I characterised as military exceptionalism), military alone, and this broader public-private manner of prosecute capability generation in our out-of-area challenges.

PR: Is the basis for this change the reliance on the technological sophistication of military kit today? If you're on a ship, it didn't take a contractor to come on board to change a circuit board: one of the WEs (Weapon Engineers) would pull one out and stick another in, and off you'd go again. The marine engineers on board would be able to change engines and parts of them because you didn't have that reliance, but have we got to a level where the sophistication of technology is so bespoke, is so boutique, is so gold plated that now it [commercial contractors] has to be part of this partnership? Have we designed ourselves a force that is now dependent on external assistance?

JL: I think the Royal Navy is a nice way of characterising this discussion perhaps. From your operational experience, Peter, you know that your guys and girls on board would be able to turn their hand to many things and keep the ship afloat operating and fighting. But now when we look at HMS *Queen Elizabeth* and we look at the strategic submarine force the number of times those ships are sailing with contractors embedded and specialists embedded from the industrial support element is significant. Completely significant. Way beyond what we would have seen in a Type 23 frigate. I think there is a sense that your instinct that we're designing technological richness and complexity into future solutions that specialists will need to support in operation must be right, just from what we've seen to date. I think what that means, though, is that maybe the nature of what we would think of as a military itself needs to change; that the force model that delivers that capability cannot be as static as perhaps it was in the past simply because the physical and the technological elements of those capabilities themselves are no longer static.

PR: If industry has made this change – and there's no doubt, I mean the vast majority of contractors that you send to sea or get deployed in the field, they go above and beyond to deliver, and it seems like industry recognised this a while ago – but the military in many ways feels like it hasn't adapted as fast at the moment to this new model. Would you agree with that? Or would you say I'm mischaracterising it?

JL: No, I don't think you are. I think a lot of our understandings of defence come from the narratives and the discourses and the stories around defence. Now, if you go back to my earlier vignette of RAF Brüggen, the last time RAF Brüggen was tactically evaluated by NATO, I think I'm correct in saying was early 1990s, just after the Wall had come down. Now, a young officer at RAF Brüggen in the early '90s is now in her or his early to mid-fifties. They are the age group of the decision makers within the militaries now, and their early career histories were very much within this traditional capability-generation stance – that of the NATO-Warsaw Pact nexus. Then of course their maturing experiences have been out of area. The militaries themselves are contested between the European model, if you like, the centre, the military exceptionalist model, military first, and this more nuanced way of trying to generate capabilities within Iraq and Afghanistan, and one point I would add, which must colour the way people think of these things, you look at the traditional post 1945 model and the West won. There's no ambiguity around that. The Warsaw Pact collapsed. We then think of our out-of-area challenges, and I don't think many reasonable people would say that the West's involvement in Iraq or Afghanistan was successful or indeed even meaningful in many ways. I think there is this challenge as well of what it means to prosecute defence and what it means to generate capabilities when, on one hand, as a young person you've been part of something that was overtly successful, and then in someone's mid-career years they've been in operations where it has been anything other.

PR: I get the point that it's changed beyond the experience of senior leaders in the military, and I'm not trying to be pro industry too much here, but it has changed for industry as well and yet they have adapted, and it feels like the military hasn't. I mean that the same guys who are fixing jets in Brüggen, working for a company back in the '80s or '90s, are now at the top of their industry game and they're the ones that have adapted to all these changes. And they will be the same people who've seen the expeditionary era and they're the same ones that trying to adapt now to the incoming investments in cyber and information and space: industry seems to be in tune with this. Is there something within the culture of the military that doesn't allow for this adaptation?

JL: I think in part we need to look at how adaptable and adaptive industry is, and it has to be, because the bottom line doesn't lie. The order book doesn't lie. Returns on investment don't lie. These are some of the most brutal performance indicators that we've devised. The balance sheet and the profit and loss account kind of lets you know how you're doing. Therefore, industry and industry executives have to be adaptive if they're going to keep their businesses alive and successful. So many businesses have gone to the wall in recent years and have exited the defence space – or have been consumed by another – because they have not been adaptable and they've not been flexible. The broad picture, as well, within industry, of course, is not one size fits all. I think if we look at the military, some of the rapid capability officers, some of the quick investments in niche technologies that the single services are making, is properly impressive and disruptive. It would be wrong, I think, to suggest that the military doesn't have characteristics of adaptability and isn't able to change. I think the opposite is true. It becomes really powerful when those behaviours and those values come together across the military and across industry. There's nothing more dynamic than seeing a military leader who wants to generate a capability through these kinds of disruptive programme or project management come up against someone in industry who believes the same and are both focused on the outcomes of their particular programme and the outputs that their contributions can make. I've witnessed it a few times, and these programmes are properly game changing. They're based on common values and exist with the military leader and with the industrial leader. I think the challenge in part is to expand upon all of that and have that culturally embedded in everything we consider around acquisition and project management.

PR: I found it really ironic that P+L (profit and loss account) makes industrial companies more adaptable than the military when the military has an existential need to be adaptable when it faces combat and to shift the way it does things, so you see the rise of the UOR, the UCR (Urgent Capability Requirements) schemes in Iraq and Afghanistan that gradually get pace behind them and that both understand. There seems to be in industry a greater ability for individuals to accept unilateral risk on decisions than you can get in the military. If a CEO decides that the future, for example, back in the 2010s was about IEDs, counter IED stuff, then they could throw money at it and go at it on their say-so. They didn't have to answer to committees or get agreement from chiefs or swap programme money. They just went out and did it. I don't know if that sort of unilateral decision making, that ability to make almost of-the-moment decisions exists within militaries anymore. Or are Western acquisition programmes so tied up in committees and programmes and approvals and scrutiny that kind of adaptability from a military perspective is now just simply not possible?

JL: I think there are a couple of things, perhaps, worth unpacking there, Peter. It's interesting to my mind, particularly in the UK, so many of our sort of leaders today have come from a Special Forces background. They've come from that background because of this problem-solving, adaptability, flexibility set of characteristics, it seems to me at least. But that is a different way of viewing this flexibility then perhaps the quick decision making that is characterised in industry. That decision making is quite formulaic itself. It is not creative and destructive and wholly innovative. It's programmatic. So even with chief executives who have reputations for taking quick and interesting decisions, they're going through investment protocols, they are having to share their thinking with their shareholders or more properly the young people who trade in their shares on behalf of capital. They're not winging it. One of the things that I find really interesting is a kind of defence industry in part that has softly generated this sense of 'we're very quick and we're very flexible', and in many ways modern industry is and it has to be. But it's still goes through this kind of rigour of investment decision making and a formality around investment decision making. I think that's good though, because I think rigour, programme management rigour are good things to want to promote. One of the things that strikes me more and more is the military, or even the Ministry of Defence, way of thinking of disruption and flexibility is different from the kind of flexibility that you'll see in well-run businesses or well-run programmes and projects. There is flexibility there; there is adaptability there; there's speed of decision making there. But it happens within a framework, and it happens within observable rigour. Quite often, the language, the taxonomy of that rigour is financial because it has to be. Of course, when you come back to talk to people in the military or in the senior civil service, very few of them hand on heart can read a balance sheet, have a meaningful understanding of investment appraisals or even profit and loss accounts, or income and expenditure accounts. That is not financially literate, and I think that's a barrier that is problematic. It's not sexy and it's not exciting. People don't join the military to spend their time looking at a balance sheet, but if they want to run complex organisations it's one of the core competencies they need to have.

PR: Given all that, John, and reflecting back on those first three paradigms that you talked about, there is a clear requirement to change many of the things we do in terms of procurement and acquisition. Even talking about acquisition of a piece of kit rather than the whole-life-cost of it is probably something that we should pass to a bygone era and we should talk more about the whole-life-cost of it and how to quantify that. I do just wonder, though, are there better models out there. Both the US and UK models for acquisitions and procurement practices get slated. Are there better models out there? Could we look to Vietnam or Georgia or Chile or Taiwan and find somewhere that has a better model that balances all these tensions?

JL: We could do. The problem with doing that, the problem with having a piece of work that identifies the best and then importing that is that it probably wouldn't align to our culture. You would have to have pieces of work in place that change the way we educated and trained, change the way we promoted, change the way we thought about holding people to account, board of governance, etc. None of that is doable within the short to mid-term. Unfortunately, we still need to generate defence capability in the short term. I'm personally not a fan of identifying the best and then sort of bringing that here, because I think that's just too difficult and too contested. What I would prefer is to look at what we have and to see how we could adapt it. Now, one of the things that is pretty clear to folk who spend time in this space is the acquisition cycle, so called, can be longer than speed of technological development. Now, that's crazy. I mean, that's really crazy and that's not fit for purpose, which means that we may have to break up that acquisition cycle. We may have to think of acquisition being not for the complete capability but acquisition for a particular technology or particular set of components, which takes us into notions around agile design, sort of open architectures and these sorts of things. That may be a better way to go than thinking about just adapting what other states are doing, because quite often even states who are supposed exemplars in this, some of the emerging powers, they often just buy kit. They often just buy capability from third parties, larger countries. That's not really what we want to do, it seems to me. I would much prefer to think about values-based procurement, values-based acquisition around sort of components rather than whole systems, agile design, technology scanning and insertion and then through life support and all of those things come together to generate some sort of whole-system capabilities.

PR: But some of those figures are going to be pretty scary. If you talked about a Queen Elizabeth class aircraft carrier, and you talked beyond the whole acquisition, you talked about the whole of through-life capability, when you're talking about the crew and the aircraft and all the associated kit that goes with it to support a platform, two platforms, two very big beautiful platforms, but trying to support those through a 50-year lifecycle, those figures are going to be – scary! This is not something that would fit a normal approvals process. That sort of approach [and the figures involved], are going to raise everyone's eyebrows.

JL: They are and being just glib for a moment, if we were to see tomorrow that the world had developed a sense that there's going to be significant disruptions that you wouldn't be able to go above a two-storey house because they'd come tumbling down as the various disruptions around the world piled in, etc., etc., you'd be a fool to design a 20-storey building. It wouldn't pass muster. Yet we have been blindly developing large, relatively slow, iconically named assets that are probably very difficult to protect and are going to be very difficult to survive in a future operating environment. I don't think that's a radical thought. I think many people think that. Something like the UK aircraft carrier programme was, for me, a fantastic target set of planning assumptions. You could have the aircraft carrier capability as a guiding construct rather than physically building it. There's nothing wrong with that. You could have that as your construct for thinking about training, development, recruitment, what force numbers you need, etc., etc. but you don't necessarily have to build it, and you don't have to build it in the way that we did. We could have thought more about having a focus on the sort of outcome, the effect, rather than just the platform but we got tied up into notions around the significance of the physical platform itself. That may be a mistake, and we may think about those things as being a mistake as we get into the more disruptive and more operating demands.

PR: John, on a final question, give our listeners some hope. Is there a feeling that somewhere down the road the plethora of new policies that are coming out or are going to come out from the government that were promised on defence industrial strategies, S&T strategies, integrated review at some stage – I mean, it's going to get better. Procurement and acquisition in the West – there's some light on the horizon somewhere, isn't there? Please?

JL: Well, I think a lot of things have changed for the better over the last few years. I think we have been able to be adaptive in terms of how we profile the programme and the projects within the equivalent programme. That's something we didn't do very well at the beginning of the century. So those things are good. I think the concern I've got always comes back to culture. It's astonishing listening to politicians, the defence secretaries, where my head is at the moment: a number of those people criticising industrialists and criticising the military have been defence ministers. They've actually been at the politically, sort of, apex of the defence operating model, and they haven't been able to effect change. The senior retired military critics were four stars, at the top of their services, and they weren't able to effect change. A number of academics who are hugely critical of this area have never served, have never been in industry. They've just written books. Quite often the same voices we hear are either overtly critical, piling into this kind of discourse of it's all broken, or they're ex senior officials, senior commanders, who weren't able to effect to change when they held those roles but somehow they know what to do now. None of that is helpful. I think the hope I have is that some of those voices start going away as they get older and older and older, and we can have a much more serious and sensible conversation around a lot of things that have been done well. At the programme and project level, you can identify exemplars invariably as based on great cultures and behaviours at the military-industrial-project management interfaces. Those should be expanded and extended and celebrated and dare I say those people should be promoted. A lot of folk, who know the answer to everything, have probably never asked the right question.

Reflections

One of those mainstays of conversations in the national security community is procurement and acquisition. Everyone, and I mean everyone, gripes and moans about it, including ministers, serving and former personnel, MPs, the media, even the National Audit Office has been known to drip like a leaky tap. But they all do so without ever offering an alternative workable solution that meets the requirements of scrutiny, value for money, oversight and that delivers output within the performance, cost and time envelope. Let's not forget that militaries are terrible customers and clients, constantly changing their minds, altering requirements midway through a build, wanting to change contract details at the drop of a hat, reducing numbers of platforms, and all for an 8 per cent profit margin that's got to make shareholders wince. Then there's the fact that the military and government expect that any profits those companies make are going to be sunk immediately into loss-leading research and development, possibly on silver bullet projects demanded by military leaders. No wonder companies have been leaving the defence sector in their droves.

Where are the procurement and acquisitions agencies in all of this? Well, it's rare to meet anyone in that line of business who is not busting a gut to try to make the whole thing work. Across NATO states, a whole variety of initiatives have been started that have short circuited the standard processes in an attempt to bring new 'disruptive' capabilities into military service faster. Yet few of these capabilities have delivered change to the battlefield in a way promised or imagined by their proponents.

Experiences on contemporary battlefields, where high intensity conflict is being waged once more, are less about the tech-heavy solutions that both industry and military leaders in 'modern' militaries have evangelised for. One can make the argument that this is simply because they are not being fielded in sufficient numbers, or used in the right way, to have the impact that they could. But it is also noteworthy that the strains and gains on today's battlefields come less from 'disruptive' and tech-heavy capabilities, and more from the traditional military hardware designed that continue to take decades to bring into service – whether a new gun, tank, missile or ship. The acquisition pathways for those are almost immaterial compared to the complexity, skills and capacity of building them.

The re-emergence of high-intensity combat in Europe, and threats of it across the globe over the coming decade, has certainly increased orders from defence industries. Yet tensions remain. For industry, there is a mistrust of a government will to deliver on a contract beyond the rhetoric: decades of military procurement programmes that have been curtailed, delayed, changed, deferred, or simply cancelled by changing governments has resulted in a cynicism from commercial actors in dealing with national governments. Well-meaning speeches by political and military leaders still are unfunded: the decision-making time from announcement to contract award is so long as to make them irrelevant.

For politicians, striking a balance between increasing spending of tax-payers money on military equipment and gaining economic benefit for the national economy more broadly usually results in a desire to build at least parts of the capability on-shore (for example, in their own country). This adds considerable time and complexity to contract negotiations and time to delivery as infrastructure and workforce up-skilling takes place. Election timetables also matter. The ideal timeline for a politician is an ability to order a capability and have it delivered within a single election session, allowing them to stand in front of it and say, 'look what we have brought to make you safer'. It is unfortunate that the build time for military equipment just doesn't happen that quickly. For countries supplying others with military equipment, there are concerns about sharing hard-won knowledge, processes and intellectual property, as well as the desire to see revenues and work for their own population.

The interaction of all these factors is mashed together in a fast-shifting geo-political world where cultures differ even when a common language is shared, and differing procurement protocols and processes are used by different state. It is, genuinely, not easy. But neither does that mean it is not possible.

And when the fighting starts, and stockpiles start to fall more rapidly than one predicted (or assumed), acquisition and procurement systems come under huge pressure to deliver. It is that moment in which systems become streamlined and, as we have seen in Ukraine, become a critical factor in determining which state will be successful in war over the course of a campaign.

9

Military Education

Each September, officers and in some states, other ranks, make their move towards colleges and academies to start a new round of professional military education. The idea of educating military leaders beyond the simple martial prowess dates back millennia. Whether the Prussians started the trend or the French, might be debated for years. Yet today, most militaries around the world see some value in preparing their commanders for future challenge by providing formal periods of education in which they can gain perspective, breadth and develop their intellectual skills. None, that I know of at least, claim to be able to equip graduates with the inspiration of the kingfisher moment or battlefield genius. But curriculums do tend to have some common elements. Some states may focus more on the financial acumen of performance rather than international affairs theory. Some on military doctrines rather than equipment and capability procurement protocols. Colleges sometimes make students do all the heavy intellectual lifting before arriving with them, giving them space to use their time in the presence of resources and professors in other ways. But these institutions are few and far between. Most set a heavy agenda of textbook learning, lectures and dissertations.

The majority of officers who go to these places to start a period of Professional Military Education, and they do so at all points in their career, usually have a large chunk of time allocated between the 10-to-20-year point in their service. At that point there are an additional set of considerations that officers face. You go on these courses thinking about how much study time you must do actual work, and how much you can allocate to family time, to pay back some of the time lost by enforced separation due to operational deployments. How many days can you spend picking up the kids from school before you return to the pressure and requirements of the next high demanding job? That's really well understood, and well reflected in a great article written by Dave Barno and Nora Bensahel back in 2019 to a class starting this process. The article is called, 'Are you enough?'[1] It's not about the imposter syndrome we all feel in starting a new job, but about making decisions on how you'll use that time in PME. They make the point that for many students, this period of education will be the only moment they have for pause, thought and clarity before the next set of jobs, sets, deployments, engagements and wars. Deciding to step back for the year and prioritise domestic over development needs to be placed in that context. But the reality is also that starting a period of PME can be daunting. Surrounded by academics in their environment, not yours, is bad enough. But then being required to write in prose and not orders formats can be utterly discombobulating.

Dr Heather Venable is from the Department of Air Power at the United States Air Command and Staff College, where she is an associate professor of Military and Security studies. She's the author of a book, '*How the Few Became the Proud: Crafting the Marine Corps Mystique*' and a managing editor of The Strategy Bridge as well as a non-resident fellow at the Marine Corps University Brute Krulak Center for Innovation and Creativity. Her snappy pieces on everything from air power to maritime strategy and on PME are beautifully written with wit, guile and insight.

What Does the Western Way of War Mean to You?

Heather Venable: I loved this Western way of war question because it's one of the reasons why I chose to actually switch from diplomatic history to military history. It's that I found more rich ideas and questions in military history and one of them was, 'What is the Western way of war?', going back to Victor Davis Hanson's pieces, and so that's one of the reasons I became a military historian. I think this is a three-part question. When I first thought about your question it would be that the Western way of war is, in many ways, very interesting because so many societies have fought at a distance and I think that the Greeks chose culturally to engage in close quarter combat. This, to them, they saw as having psychological advantages but

---

[1] David Barno and Nora Bensahel, 'Are you enough?Our speech to the PME class of 2019'. *War on the Rocks*, 18 September 2018. https://warontherocks.com/2018/09/are-you-enough-our-speech-to-the-pme-class-of-2019/.

I think in many ways this also neglected the psychological costs of fighting at close quarters and it really privileged the idea of a decisive battle and we still think about, in our influence by the first type of Western way of war, the myths that we tell ourselves about Greek warfare and how in some ways they continue to idealise portions of it. Even though we're idealising the myths rather than the actual reality of history itself. Then I think it transitioned, for me at least, into a selective interpretation again of Napoleonic history and this continuing quest for a decisive battle which is understandable, but something we wrestle a lot with at the staff colleges.

Then in World War One, and even before, there was a deadlock because of the changing balance between offensive and defensive weapons, which has a lot of implications for today as well. A lot of people came up with manoeuvre warfare as a solution and I think manoeuvre warfare is a fascinating idea in theory; I'm not sure how often it works in practice. I think often when it does work, one of the key factors when it does work is because enemy morale often is low. But what happens when enemy morale isn't low? and so I think history is very useful here for going back and trying to check whether manoeuvre warfare actually works in practice. I have written about that and I am continuing to write about that and explore the concept.[2] Then finally I think that the new Western way of war that we've seen maybe since Operation Desert Storm where air power also really comes into the picture is that ironically we've reversed that trajectory of embracing close order combat and now increasingly have tried to fight from a distance and in that we've also added this humanitarian element of trying to avoid casualties. So that's, to me, how I see the progression of the Western way of war. As a historian, of course we don't like generalisations and we always struggle between trying to give people useful ways to dissect huge periods of time and the reality that there are no easy answers for military practitioners.

Peter Roberts: Heather, that was an amazing answer, it was multi-faceted, there's loads there that I want to dig into but your last point about airpower has always struck me as quite interesting because I was having a

---

[2] David Alman and Heather Venable, 'Bending the Principle of Mass: Why That Approach No Longer Works for AirPower'. *War on the Rocks*, 15 September 2020, https://warontherocks.com/2020/09/bending-the-principle-of-mass-why-that-approach-no-longer-works-for-airpower/; Heather Venable, 'Decisive Maneuver is the Army Equivalent of the Air Force's Historical Emphasis on Strategic Attack as "The" Answer', *Linkedin*, 3 November 2023. https://www.linkedin.com/pulse/decisive-maneuver-army-equivalent-air-forces-emphasis-heather-venable-fzkhf/.

conversation with a mate of mine the other day and we were talking about the experience of Afghanistan is, in many ways, that it was about using airpower almost as a sniper rifle. This is the way that we had developed our ability to use airpower and it seems a really expensive and wasteful way to use such an incredible military asset. We don't seem to have developed more original theories despite all the time we've invested on it. It feels like airpower doctrine has just stagnated; it hasn't moved forward. Do you think that's true, or do you think there is some change on the horizon? Or are we stuck in this idea that CAS (Close Air Support) and BAI (Battlefield Air Interdiction) against an individual combatant is the way ahead?

HV: That's a really tough question to answer in so many ways. I think that the air force really struggles between trying to advocate for [an argument summed up as] "there's more to air power than just serving as flying artillery". and when you use it that way it's a huge costly endeavour and surely there are more efficient ways. But on the other hand, with the casualty aversion [of modern Western military operations] in trying to avoid casualties on the ground and trying to be a part of the joint fight and supporting and getting the over airpower's reputation of too much being go it alone, it has a really hard time saying, 'No this isn't an efficient way.' One of the main things I like to talk about a lot is that we've had the luxury of being able to throw a lot of airpower at the problem. But in the future, in terms of thinking about Great Power Competition, what kind of choices will we have to make about employing airpower? I know the army, the US army, has anticipated this and is trying to develop more long-range weapons, for example. But then you have the problem that now all the services are developing potentially redundant resources and how do you pick and choose? That's why you see a lot of the food fights that are picking up in the US between the three services. I'm sure you see similar things over in Britain as well.

PR: Yes to the sort of budgetary fights, but for me, we're losing the rigour, this intellectual depth that gives us an idea about how we should be fighting with these tools. That's why, because no one's come up with a better idea, everyone's inventing the same things, albeit in slightly different ways. As a result I'm sort of struck by this idea that through all the thinking, all the teaching, everything we do about whether it's airpower or anything else, we somehow have failed to progress tactically and doctrinally, as well as operationally and conceptually. Even Multi Domain Battle and Multi Domain Operations, aren't both these ideas just a way to use things more efficiently? It's not a battle winning technique and yet, if you looked for success of all the graduates, all the PhDs, all the money and time we've invested in PME, we haven't developed commanders who can win wars. What we've got is losses all over the place. Since 2000 we've continually lost engagements. People have been heroic and brave, don't let me take that away from them, but the fact is we've lost, tactically, operationally and strategically. Part of this must come down to our way of thinking about, not just airpower, but how we educate our people. Do you think that's true?

HV: I do but I would like to circle back first to the multi domain piece because it's something I've written a lot about and feel strongly about. I think that the Air Force in particular may have some of the best tacticians in world history, maybe I'm drinking a little bit too much blue KoolAid but I think there is an argument that the students I get in my classroom are tactical experts. But then the problem comes at Air Command and Staff College, our role is to educate them at the operational and strategic level of war. That requires a very different mentality because for operational and strategic [levels of war], you're getting more into theory, doctoring strategy and those change less over time and so history, in many ways, becomes more useful for thinking about because you have more rigorous thought and examination of the past. For example, two weeks ago I taught OEF, or Operation Enduring Freedom, using a publication that ended in 2005 and it was disconcerting for the students and myself to teach it then and I could only imagine how much more it would be now mid-August after the events of this past week.[3] We really struggled between, 'Well how do we talk about small war or COIN or whatever term you want to use when there's very little written, especially recently, that's rigorous because we can't get into the archives.' This is something I feel strongly about that after Operation Desert Storm, the Air Force quickly released a multi volume Gulf War airpower survey that had some very reputable non-KoolAid drinkers doing the study.

In many ways their findings went against some of the louder voices in the Air Force that were more zealous about how airpower should be used, finding that it was really successful in the Kuwaiti theatre of operations

---

[3] The interview was recorded the week after the Western led withdrawal from Afghanistan occurred, and the widely reported 'Fall of Kabul" to Taliban forces. See the White House official statement of 31 August 2021. https://www.whitehouse.gov/briefing-room/speeches-remarks/2021/08/31/remarks-by-president-biden-on-the-end-of-the-war-in-afghanistan/.

and not necessarily as a purely strategic campaign. But yes, we don't have anything like that since then and so, the US army released a couple of years ago almost 1,000 documents online that you can access through the War College and there are some things that are redacted but still, you can get a lot out of that. There's nothing for the Air Force, and I think that until we start wrestling and can study more about what those conflicts are, it's hard to feel like we're being honest with ourselves and can learn from what happened. Especially when you get from that early phase of the conventional conflict against Baghdad that lasted about five weeks in 2003, and then suddenly turned towards the unconventional. What I want to understand is; what were airmen thinking? What kind of different orders and ideas did they have when they were trying to figure out how things that are going differently, not like we expected? what do we do now and how did they adapt? We just can't get into that because the Air Force won't give us those records.

PR: It's amazing, isn't it. I mean I'm a big fan of The Spear, the Modern Warfare Institute podcast, where it's their front-line reflections of people who've come back from combat and they talk through their engagements and you get people who talk about, whether SF [Special Forces] operators or an Apache pilot, you get some real feeling for what they experienced. The one thing that's always been missing in there was a reflection on airpower. Whether it's fast air or strategic lift, there's a bit of rotary wing aviation in there, but you just don't get the feel that that [fixed wing combat power] knowledge is being shared, that that experience is being shared. As you say, it's not just the written records but it's the oral histories that really seem to be missing from the Air Force at the moment. I don't know if that's something that is culturally inside all air forces, because it doesn't happen in Europe either.

HV: I'm not sure, and I think that when you're talking about individual people and their memories, I was just having this conversation, I had a PhD [student] who was also a Lieutenant Colonel at the time. He gave a lecture for us two years ago, or so, on Operation Inherent Resolve and then I had a student come up to me and the student said, 'Well that's not how it occurred.' Having a little bit more experience I would give her a better answer now, but everything was from open-source material. So to say that everything in that lecture was presented incorrectly revealed a very myopic perspective that this person assumed that she knew the real story. Also just that the one piece, the US led piece, and this comes I guess back from some of my graduate studies in the British Empire and interest in colonialism and agency, that we have the piece of the story and if we can just figure out what we're doing, everything will flow. I think that one of the problems with these conflicts is that we've seen our side but now fully wrestled with the Clausewitzian triangle with the people element and even with all our technology, the people element is still critical.

PR: I want to come back on that point: relating exactly what you're talking about – the contemporary records, the contemporary experience. Students will go to their PME courses emotionally invested in recent campaigns. Seriously invested in them. Which makes it very difficult to examine those with any sense of perspective. Most people will understand that it is impossible to look at these military campaigns with a dispassionate eye when you have been in the thick of the fighting and going back in history provides an ability to do that in a slightly different way. I guess, is that one of the secrets of success, is to remove yourself emotionally from the curriculum that you're looking at?

HV: So those are really great questions. I think that my first piece of the answer would be I think it's true but it might be hearsay. But even if it is just hearsay it's great hearsay. It's what the [US] Naval War College said when they were trying to figure out how to study the Vietnam War – they said, 'Students who feel this too personally can't study this.' So that's supposedly when they added Thucydides into the curriculum as a way to get at the same issues. So it is very much true for us that we get to our modern airpower course and there is a lot of emotional investment and a lot of, I've missed my kids birthdays, I lost friends, all of these things that go into understanding and their feelings and so it's hard to think about this dispassionately. In my last class, honestly, my students didn't want to talk about Afghanistan, they didn't want to talk about Iraq because they said, 'That's what we've been doing for so long.' It's interesting because in an educating airpower book that came out in the last year, it's a collection of airman's thoughts or airmen professional military education in Italy, the US, Canada and Britain. One of the starting pieces of analysis that the editors offer in the beginning is that airpower education has become too conservative. I don't know whether that's true or not but I think it's interesting and as someone who's responsible for a piece of the curriculum at ACSC, I want to consider that. But how do you draw the line between traditional critical thought and cultivating creativity which you don't have the necessarily look at the newest, you can look at ancient case studies.

Because you're cultivating a way of thought more than providing tactical knowledge. Because what we're doing here changes less over time, because we're interested in theory, strategy, where when you get into Tactics, Techniques and Procedures (TTPs) those are the things that change more [dynamically]. So with the operational and strategic level of war, I think we do have the advantage of more stability so we can look more deeply into things. On the one hand, I think we can get more out of focusing on Vietnam than Afghanistan because we have a better understanding of what happened and a bunch of different perspectives that we can bring in whereas for the other reasons I've already given about Afghanistan and Iraq, it's a lot harder. The other challenge though is avoiding being too conservative, thinking about how you look into the future. This is something that's not necessarily my responsibility because for my course, we end [our examination of the history of airpower] in 1973. But how much artificial intelligence do you include and why, how much robotics, how many swarms? Because, and this goes back to our earlier discussion in terms of multi domain operations or whatever it's called today, it changes every day so it's hard to keep up. But I think if you ask students what the answer is and what's the future of airpower, they're going to say things like, 'Swarms, neural networks and AI flown jets,' and all these things that we see and these are all things. But then we stick them on a proverbial airpower dartboard of strategy, instead of thinking going back to the foundational ideas and seeing how we might actually try them.

This gets into a different conversation that I have wrestled with and don't know the answer to at all, which is when you're thinking about teaching science fiction, and some of the futuristic novels. I'm not personally convinced how useful those are, although I know a lot of people who think they are. I feel like when you get stuck in a way of thinking that cuts off some of your thinking – because now you have such a powerful vision and you've kind of seen this in the space force (it's really hard for the space force to get away from Star Trek and Star Wars influences), although they also claim that they came up with it first. But these visions are so powerful in shaping things of where the future's going. This is a very long-winded answer to say it's very complicated to try to figure out how to keep airpower and professional military education from getting stagnant and where we draw the line and what issues we consider when we're approaching it.

PR: Across everything we've talked about is the idea that a lot of us are stuck in a conceptual envelope of our own making that constrains us.

Whether that's airpower, whether it's the idea that we regard the future solely through the eyes of P.W. Singer and August Cole. Whether we think about the way of warfare as counter terrorist operations and nation building and how useful is that thinking when we talk about Great Power rivalry? How much can we learn from Afghanistan, Iraq or even Syria and Yemen that applies to the next era? I think this, for me, has always been the point about PME: it should be about learning how to stop thinking the way that you currently think [as a tactician perhaps] and to not be constrained by the resources that may be given around you. To break out from your own experiences and to start to learn how to tap into the experience of others. I know that might be history, that might be other campaigns, but it might also be writing by the Russians or the Chinese who are actually doing some interesting stuff on airpower right now in doctrinal terms. I just think that in a way, we've become very stuck in a technical, scientific examination of our professional military education and what we're lacking is the artistic part that grows thinkers. We grow people who solutionize every problem, but they solutionize it with a set mindset. I don't know if you'd agree with that or you think I'm talking rubbish?

HV: No, the only thing I would say is that from my experience at my college is that we're not very technical at all and we are more arts based, we teach war as an art more than a science. I think that because we're teaching majors who have about 14 years of experience, that this gets to a question that I've been thinking about a lot this week which is whether the war colleges have more emphasis on the college part at the exclusion of war or if they need to put more war and real military stuff back in. I think that, as again someone who's responsible for determining parts of a curriculum at a staff college, it's a question I take very seriously and give a lot of thought to and don't want to close off my mind to. But I think that the argument that we saw on social media this week was that the Taliban, in many ways, had succeeded perhaps because they were these manoeuvre artists, that they were able to shock the opposition into freezing up and not being able to respond. But I think another solution would be also to say that they were political masters and that they understood how to achieve their objectives by more political and non-kinetic means in some ways in this case. I think that's a lesson for PME. Then again, we have this ongoing battle where we try to figure out how we teach kinetics, but also non-kinetics. That we recognise have always been important but are increasingly becoming more important and we see that with the Taliban's foundation that they led, for at least short-term success, recently.

Probably, unfortunately, longer term success. I like to always remind my students that Clausewitz said war is never final (and that's probably one of my favourite lines from Clausewitz). I think that this week, mid-August shows us that the War College has to be a place that is a proper mix of war and college because people are senior officers that are going to be generals need to understand politics, they need to understand the economy and they already have been studying military history and have been practising it for all of their careers. Now that they're going to be engaging with civilians, that requires them to have more of a broadening education. Here at the staff college level, where we have mid-career officers, we are still more focused on war and joint war fighting because in the coming fight we are not going to need people who have to be conversant in politics, cultures and economics; all those things that our previous students, our graduates, found themselves doing in Afghanistan. When you're thinking about Great Power competition you have to think about it not just as a peer-on-peer fight – which must be taken very seriously and for which they must be adequately prepared for – but also a less conventional fight that our opponents might chose to pick a fight and we might find ourselves in. Either way, we have to think about using the military in a larger pursuit of political objectives and that requires the college part and the war part.

PR: When you're talking about 'the Taliban being great manoeuvrists', thgis sort of comment is symptomatic of the temptation for military professionals looking at any conflict to apply the lessons that they want to have: to selectively use evidence based on limited, skewed evidence. That might be from the Taliban in Afghanistan (as manoeuvrists), its might be the Assad regime in Syria, the Houthi's in Yemen, the second Naguro Karabakt or whatever we were talking about: there is a military predilection to make those conflicts conform to a desired narrative. So in looking at some of those fights, we hear a lot of commentary that says, "This is about drones and the futility of armour", or for the IDF Gaza conflict where everyone was taking about AI and IAMD:[4] There was an almost centrally driven Western narrative that said, "These are the big lessons, do not deviate from them". I think in some cases we shouldn't allow militaries to draw the lessons

---

[4] Artificial Intelligence and Integrated Air and Missile Defence. See, for example, *The Economist*, "How Ukraine is using AI to fight Russia", 8 April 2024, https://www.economist.com/science-and-technology/2024/04/08/how-ukraine-is-using-ai-to-fight-russia, and *The Economist*, "How Missiles are changing the Middle East", 22 November 2023, https://www.economist.com/films/2023/11/22/how-missiles-are-changing-the-middle-east.

because they do so in order to impose their own mental frameworks over the top of them. Which is why this idea of giving them an extra skill set [in terms of the politics element] is so important, I think you're absolutely right. But in that it must be really hard to break some of these conceptual shells for your students, these mental constructs for the use of military power that they arrive at college with. Do you smash them apart quickly, do you deconstruct them slowly and methodically? How do you go about breaking those down? Is it a very individual thing or is there a generalist way you can go at it that works for everyone? For me, it's the military, so I would go at it with a sledgehammer...

HV: One of that challenges that we face is that many of our officers that we teach have been selected because they have excelled at being officers and they've excelled at being officers by having been given questions and they've answered them in excellent ways. So a lot of us here try to get them out of that mindset and to think not about how to answer questions but how to ask them and we do exercise with cognitive dissonance and how and why ideas come to be held so tightly. That goes to what you were saying with some of the examples. Again, it goes back to our, 'Well what do we study about recent conflicts and how do we necessarily know that we're concluding the correct things when we only have a short perspective' and we don't really understand fully why situations played out like they did. We can speculate and then we tend to, as your examples show, hone in on platforms and capabilities and technologies without thinking about all the things that make those capabilities work. The ideas that underpin them, the people, the training, everything else and we just make sweeping conclusions and stuff.

PR: My final question to you, is that I want to know what the best bit of advice you could give a student who's starting a period of professional military education is? What's the one thing that you think is the secret to success?

HV: I'm really interested in creativity and how the brain works and how ideas come out, and white space. I know my best ideas occur when I wake up too early and I'm lying in bed – it's 4am in the morning and I don't want to get out of bed because I'm too lazy and so there's nothing to do but just to sit there and think and that's when I get many of my best ideas. I tell students walking around, running, all these daily things when you're not consciously thinking, that's when your best ideas are going to hit you. So white space, I think, is really important in PME and sometimes we're guilty here of filling that space up and trying to make the most of every precious thing, you can be thinking while you're playing golf. I hope no one will fire me for saying that. But I do think students should read but I think also, they need to read and I would hate for them to think that this is completely a year for relaxation. But I do see it as a year to recharge. You should recharge with your families, you should recharge your mind, most of all. That is really challenging because for students that are highly motivated and want to learn: we give them all these books and they have so many details to ingest, and we are asking them to, first of all, learn Air Force history, and doctrine etc. But that's not really the primary reason that they're here. I think the best way I know of to recharge your mind and prepare for the future is to learn how to deconstruct arguments, to engage critically with ideas to make arguments of their own.

To do that you really have to read actively so I try to get students to not take notes. Instead, I encourage them to just write what they remember [from a lecture or a book]; that helps you get at the essence of what's important. I try to bring in more active notetaking and so we can use pictures of circles and that way you can look through the chapter and find out what really matters. The students here subdivide their notes and they have one student who will take notes for page one through ten, and the next student will take notes from ten through twenty and then what you get is this volcano of material and I think one of the most important challenges, especially for officers in 2021 when we have way too much information, is how do we pick and choose what's most important. If you're using a division of labour to read and take notes then you're not refining that skill, and that's a skill that we really need today.

PR: I couldn't agree more, Heather. There's just so much information out there distilling the essence of it and knowing where to look and what to read, to believe and to value is critical.

Reflections

Whether on a service specific course, or one of the more generic 'joint' staff college courses, there is considerable overlap in the syllabus of most PME courses. With the exception of a handful of military colleges, the vast majority of students are required to focus on a narrow seam of learning that institutionalises prior preconceptions. Instead of allowing students to utilise the time to broaden their knowledge and gain a better understanding of war (and warfare), military chiefs and military educators have determined that courses are best utilised in generating the skills of accountants and mediocre tacticians. It is peculiar that despite the rhetoric, that senior leaders in Western states do not want officers who can think independently, widely, critically, or deeply about bigger issues, preferring instead to perpetuate a system of professional education that advances the failures of previous generations. There are some notable exceptions; courses that provide the challenge and freedoms officers need to develop genuinely divergent intellectual skills, alongside the confidence to challenge and learn.

There is a notable tension in PME courses, particularly with students returning from operations, that reflects many of the conversations about contemporary militaries. What lessons to learn from current campaigns that will have relevance and utility in future warfare. The disregard of lessons from history, in favour of those gained from the most recent conflict, is a trend that has reversed the processes in place in PME prior to the 1990s. This tension reflects the classic juxtaposition of 'a war'/'the war'; a dilemma that all militaries have faced across millennia and few have solved.

10

First Order Questions

The public discussions over the Western way of war tend to focus at the military tactical and operational levels, only occasionally referencing the political aspects (perhaps to lay blame at their door) and elevating strategic challenges to an abstract position such that their answers do little to assist in developing more coherence and utility for the future. Yet there are critical questions which must be addressed before a state commits to military action. Per Carl von Clausewitz, "The first, the supreme, the most far-reaching act of judgement that the statesman and the commander has to make is to establish by that test the kind of war on which they are embarking on. Either mistaking it for, not trying to turn it into, something that is alien to its nature. This is the first of all strategic questions and the most comprehensive."[1] This idea, that we should expect more intellectual engagement and honesty from our leaders (both military and political) has particular validity when we try to understand why nations would go to war. In making those decisions – as a choice rather than as an existential requirement – leaders in Western states have relied (at least in recent history) on a myriad of less-than clear (or truthful) rationale.

It has been argued that European nations – whether NATO members or not – engaged in US led wars and campaigns of the 21st Century in order to secure US security guarantees in the longer term; to be seen as 'team players' in a world order dominated by the US. The contribution of very public international political support but small force rations allowed America's partners to freeload on the US taxpayer for their defence and security. But in doing so, many European states failed, dramatically, to understand or pose the questions required of them: a series of inactions that

---

[1] Carl von Clausewitz, *On War*.

failed their own populations and their partners. This critique applies just as well to interventions in Libya, Mali, or elsewhere, as it does to campaigns in Iraq or Afghanistan.

The senior levels of national militaries were also at considerable fault; settling for vacuous statements about 'end states', 'exit strategies', and dead dogma instead of seeking answers to the first order questions, like what the political purpose of the war is. In turn, political leaders failed to demand of their military chiefs' explanations about what would be required to 'win', to achieve their ambitions, and why the plan (as laid out to them) would work. The simple acceptance of a military plan – often shoehorning coalition forces into an American set of PowerPoint slides – sleepwalked partner nations into a myriad of campaigns, the consequences of which they completely failed to understand.

Still today, little discussion is given over to the nature of differing wars, the nature of the societies being fought, the dynamics of a learning adversary, the diametrically opposed requirements – perhaps – for the complete destruction of an adversary rather than a more practiced decapitation of leadership.

Such topics, and the education of political leaders (even if it is limited to allow them to understand which questions to ask of the military and intelligence community), have been absent from Western cultures (certainly in Europe) for at least 40 years. The shadows of both World War One and the Cold War, with their requirements for deep consideration of unpleasant but necessary scenarios, have been forgotten. As John Stuart Mill put it, "Both teachers and learners go to sleep at their post, as soon as there is no enemy in the field."[2]

Eliot Cohen was the Robert E. Osgood Professor at the Johns Hopkins University School of Advanced International Studies where he taught between 1990 and 2021. He served as Dean there from 1990. Cohen received his BA and his PhD degrees from Harvard and after teaching there and at the Naval War College founded the strategic studies programme at SAIS. His books include *The Big Stick*, *Conquered into Liberty* and the one probably most familiar to our listeners, *Supreme Command*. In addition to public service in the U.S. Department of Defense, he served as a counsellor to the Department of State from 2007 to 2009. In a speech about "The slumber of decided Opinion (against the tyranny of the majority)", Eliot made a

---

[2]   John Stuart Mill, *On Liberty* (John Parker and Son, 1859).

compelling argument about what needed to change. It was the trigger for this discussion.

What Does the Western Way of War Mean to You?

Eliot Cohen: I'm not entirely sure that there is a Western way of war which has been with us since the Hoplites. I do think that at different times over the last centuries it has meant some things that are very distinctive. At the moment what I would say is that the Western way of war is distinguished by a level of scrupulousness about civilian casualties, about precision, about legality, which is way beyond the norm. If you look at our past, has the Western way of war been completely ruthless and destructive? Yes, but so has many other cultures' ways of war. Has it been purposive? Yes, some of the time: but I think, once again, other cultures have used war in very purposive fashions. So, I think you really have to situate it in a particular piece of the continuum of history and at the moment the key distinguishing feature is that level of scrupulousness, which is, maybe, in one way the right thing and another way a tremendous source of weakness.

Peter Roberts: Yes, it's interesting: Politically it's very hard to define a Western way of war almost philosophically as so different from either an Eastern way of war or a Russian way of war, or even a Prussian way of war. Militarily you can divide it up tactically as, you said, the use of precision weapons or manoeuvre or fire power, whatever it is, but this idea that more recently we've come to define a Western way of war around a set of morals or ethics, or values that we haven't done previously is something that is genuinely quite new. Of course that approach is very marketable; it's hugely fashionable and you can make movies about it. There is, perhaps an almost biblical tone to the whole thing but not much evidence of its use by our ancestors. It doesn't have a long historical legacy attached to it.

EC: No and I'm not even sure it's biblical because when you look at the outcome of most wars in the Bible, they usually end up with a mass extermination of one kind or another. Part of this is simply that technology creates a certain set of possibilities for us which never used to be the case, but I also think that Western societies have changed and a way of war reflects the nature of the society in some measure but that's never really particularly fixed. We're still the same people who after all did Dresden and

Tokyo and Hiroshima. So, it is the product of a particular moment but we're dealing with adversaries who I think don't have quite those compunctions.

PR: In many ways it feels like we've changed but they haven't. If we go back to the Cold War in which – we should recall – there were less ethical and moral concerns when we thought about the fighting that was going to be associated with tactical nuclear weapons, with chemical biological weapons, which we foresaw as part of the great battle across mainland Europe. This was predictive, we trained so much in chemical, biological and radiological techniques of how we would continue fighting that it was largely preordained in many ways for us to expect this to happen, to train against it and to deliver against it and yet it's our expectation that has changed about what we think war should look like, perhaps those competitors, rivals, whatever we call them, the others, don't necessarily feel that way.

EC: Well, part of this too it seems to me is the result of the long shadow of World War Two. In the immediate aftermath of World War Two people can certainly envision war being as terrible or – if that was possible for those who experienced it – even more terrible than that conflict. But we really are past the shadow of World War Two insofar as countries like the United States or Great Britain are concerned. That's partly because we've been living in such an era of security that it doesn't really occur to us that you might have to do something remotely as devastating as you did during the Second World War. Even when after 9/11 you had a substantial amount of violence inflicted on the United States, there still wasn't the necessity, although I suspect there would have been the will of levelling cities, there was a lot of determination. However, now you could do it in a way that was pretty precise. I think one of the things that may have also slipped away as the shadow of World War Two failed, which may not be the case with other cultures and civilisations, is an awareness that, at some level, what the wars of the 20th Century were about in many ways was breaking the will of the opposing society and not just defeating its armed forces in the field. That does remain a kind of fundamental truth about warfare which we would prefer not to think about, but I suspect is still valid. If ever there were to be another substantial conflict would very much be with us.

PR: It also feels like perhaps that society has developed these scruples because society as a whole hasn't really experienced warfare in the recent past and yet if you go to Aleppo or Raqqa and you see the devastation that is hugely reminiscent of Dresden or London during the Blitz or elsewhere, there are some real shifts in how people feel about warfare when it's a lived experience. Very few in the West have had that experience brought home to them, there's only so much you can depict through news clips and the occasional film. If, as you say – and I agree wholeheartedly with you – our societies have changed, liberalised particularly over the topic of war and warfare, and specifically the conduct of combat operations because of our experiences in undertaking military actions in an era of peace and security, do you think those attitudes [the moral and ethical behaviours we aspire to and bound ourselves with] could revert to something more base?

EC: Quite easily. For so many of these things the Israelis are a really interesting case study because that is a military that has shifted very much to a world of what the Israeli Chief of Staff, who's a former student of mine actually, Aviv Kochavi, calls industrialised precision. I mean, they really conduct precision warfare but even there I think you can say that they can afford to be, and there are a lot of tactical advantages to precision, of course, but even they I think are still to some extent shielded from the reality of a war in which their homeland is really under massive attack, where you're taking not a couple of dozen civilian casualties, but thousands. If, which obviously we hope didn't happen, you had an Israeli-Hezbollah war where Hezbollah by some counts has over 100,000 rockets stored away and they're unleashing thousands every day at Israeli cities, and Israeli civilians were dying in not the hundreds but the thousands, I think you'd very quickly see levels of violence unleashed which were much more reminiscent of what we saw in World War Two and the aim would shift to break the societal will as well as eliminating enemy forces.

PR: It's fascinating to me that leaders don't necessarily reflect society, not necessarily military leaders but political ones in this debate, right? So, it feels like political leaders are more in touch with the contemporary warfare than perhaps societies. They're willing to consider means that even military leaders won't think about in some ways because not that they have flashes of insight, but they get the sort of long-term political ends and the requirements that might be required to get there. Do you think that's true?

EC: I think it depends on who exactly it is that you're talking about, but there are flashes of that. Somebody who does not always have a great reputation as a Commander-in-Chief is George Bush; I remember during the darkest time in Iraq when military leaders were quite willing to think about disengaging and they were, in particular, talking about extending deployments to Iraq. They said, 'Well, it could break the force if we don't begin drawing down.' He [President Bush] said, 'What will really break the force is losing a substantial war.' There was a certain kind of human insight there and he was right and it was a serious strain on people to have tours extended from twelve months to fourteen months. But I think having seen our forces in the field during that period you could tell that there was a profound morale difference as a result of the fact that they thought things were turning around as a result of the surge. So, it's true and political leaders do have a certain kind of insight that military leaders don't always have.

PR: Some of them [politicians] certainly have a different perspective. I do not think we found enough military leaders at the time to pose that question at that moment. There wasn't even a thought about, 'What would a loss mean?' That wasn't the primary or even secondary factor in people's minds. Perhaps that's what many Western states have almost forgotten; that interplay between the political masters and the military leaders. There is an instance during that period when the then British Prime Minister David Cameron said to his generals, 'You do the fighting, I'll do the thinking.' There was a great deal of pushback about that but politicians should be bringing something to the fight. It's not just their part in constructing a 'grand' strategy, it's not simply about the political apportionment of national resources, it's a different way of thinking about the conduct of war and warfare.

EC: I think one of the things that they should be bringing is asking the first order questions, particularly, and again I'll go back to the idea of the long shadow of the Second World War. The big difference between our armed forces today, I would argue, and those of the '30s and '40s was that they've been massively institutionalised ever since then. They have their ways of doing business. I'll give you just a small example of this. The United States military got used to the idea of rotating headquarters, rotating entire units, brigades and above into theatres of conflict. That's just what they did because that's what they had always done. It took political leaders to begin asking, I wouldn't say necessarily particularly effectively, to at least ask the right question which is, 'Well, why are you doing that? Maybe it makes sense to rotate battalions but why rotate a divisional headquarters?' It's not like these are small groups of men and women going into combat all the time who need that kind of cohesion and the result was the loss of continuity and so on. I think that's a very important thing for us to remember that when you have these well-established bureaucracies that are used to operating in a certain way, sometimes it does take a politician to say, 'Well, hang on a second. Why exactly are we doing that?' Really, the question can be that simple.

PR: I had a conversation with a former Secretary of State in the UK who said, 'It's all very well you saying this sort of thing about how we should question them [the generals] but do you know what? It's quite hard to go into a room full of military leaders who between them have got more than a century of military experience, albeit in the wrong sort of wars and you know it's the wrong sort of war, but they have their uniforms and their medals, and their combat stripes and their deployments and everything else under their belt. All the technical knowledge and for little old me to turn around to them and go, "I don't think this is right," it takes quite a leap sometimes.'

EC: Yes. Look, it is, and I remember I had two sessions with President Bush on Iraq strategy and the first one in retrospect I whiffed. The second one I was actually considerably more straightforward but that's because my son was about to be deployed in an infantry brigade and so I was feeling chippier – but it's very hard when you're dealing with any kind of authority. I think the way to do it is to ask the very basic kinds of questions and not be put off by answers that are not convincing. That, it seems to me, is the most effective technique for civilians to actually begin to get things changed. One of the things that happens when you do that, of course, is one quickly discovers that the military is not a monolith and there may be some colonel who's a backbencher or some disgruntled two-star who actually has a different view, and that's by the way another thing that civilians can do. I mean, it can drive a certain kind of military leader crazy but don't be overly respectful of the chain of command and assume that the four-star is smarter than the one-star because they, of course, frequently are not. You see those games being played where people are playing generals off each other. We had talked a little bit about what kind of questions one might ask, one is the [General David] Petraeus question, 'Tell me how this ends, what are we trying to achieve?'

For me the most important question which I have an occasion to ask a few times was tell me why we think this will work? So much of government to include in the waging of war is about inputs: How many troops? How many diplomats? How much aid money? How many munitions delivered? How many targets struck? I think one of the things that civilians can very usefully do is just pose the question saying, 'Okay, I understand that we're providing all these inputs. What is the story that links the inputs to the outputs?' I'll just say one other thing about that. Amazon has been an extraordinarily successful company, one of the things that has struck me in everything that I've read about Amazon is that Jeff Bezos always insisted on if there's a new product or new initiative that it be framed as a story, not as a bunch of PowerPoints. I think that's probably the most effective way of getting people to think strategically. Say, 'Okay, tell me what the story is.'

PR: I assume that a lot of it depends though on the customer receiving the story too. I was struck when we were talking about those big questions – like, 'how do we think this will end' – by memories of the various military plans I've seen at staff colleges and wargames, exercises, and so forth. The plans that the Blue Force comes up with are invariably adversary agnostic. They don't look at warfare as a dynamic, they look at it as a one-way thing: we impose these conditions on the adversary and they react. In developing this form of thinking – and this is a particular bugbear of mine – staff usually end up so wedded to the perfection of the plan that they develop their metrics for success based on what the Blue Force is doing, not on the interaction or dynamics with the adversary. Those developing the plans, from the 'infallible' military leader downwards, have a mindset that has predetermined what will work out, agnostic of both the adversary and the publics. The academic, scholarly scrutiny behind this is the theory of change. What's your theory of change [Is a question that is rarely posed however]? It's 'How does this work'? Then, 'Why does it work'? These questions are so critical.

EC: Let me tie that back to your opening question about the Western way of war. I said that I think it's really very much shaped by the time that we're living in. We talked earlier about the shadow of World War Two. What I think we're seeing there is also the shadow of the Cold War where you had this frozen conflict where staff colleges, particularly in the United States and the UK, were used to a very well-defined kind of opponent and a very well-defined sort of circumstance and thank God it never really went to war. We ended up fighting some real wars which were viewed as anomalous and that's how we ended up I think engaging in military planning against generic opponents as opposed to real opponents. That I believe will be a mistake for the future. The thing that's unfortunate that reinforces that is to openly speak about particular opponents these days is to risk various kinds of political kerfuffles which can range from embarrassing to serious diplomatic incident. So, people go pretty generic but, again, my own military experience was minimal but I was a reservist in the 1980s and all of our training wasn't against the Soviets, it was the OpFor, the opposing forces.

There was nothing that made you think that they were actually thinking like Russians, I mean that you would learn what was essentially Soviet doctrine and so on but that was really more in templating, for example, where are they going to put the divisional artillery group and things of that kind. It wasn't actually training you to think about a real-life reactive adversary.

PR: It is quite peculiar: we know the theory about fighting the commander and not the force, don't fight the doctrine, fight the way the commander will fight. We have the requisite information about the adversaries we'll be facing but so rarely do we put that mindset in the red cell in exercises and war games. We don't say to [the actors playing] an opposing force, 'So, you're the red team for this one. You need to act like Gerasimov', or whoever you picked that you want to pull out. We never exercised against the mind of Soleimani. Do you think that's one of our greatest failings?

EC: I think it is a failing in that we're not preparing people to think that way. In a staff college you're obviously preparing people for an uncertain world but there are ways that you can do it by historical simulations and so forth. Again, the Israelis are quite interesting as a kind of Western laboratory. They do think about particular opponents, they think about Nasrallah in Lebanon. They certainly did think about Soleimani. They think about the particular leaders that they're dealing with in Gaza. Now, of course, that actually opens up another thing that's now part of the Western way of war which is going after individuals: but that isn't just going to be the Western way of war, everybody is going to be targeting individuals. That's a whole new dimension to war really, that putting a premium on taking out particular people.

PR: That all stems from this belief in [JFC Fuller's] Plan 1919, doesn't it? The Western personalisation of war is an extension of beheading the snake, a decapitation strategy that eliminates the leadership and [a belief that] everything else falls away. Yet from our experiences in a post-Cold War world, we've seen that decapitation strategies are not a very successful way of taking a war to an adversary in many circumstances.

EC: No, but it can feel very fulfilling. For example, the United States orchestrated the shoot down of Admiral Yamamoto, but that didn't really shape the outcome of the Pacific War. Again, part of what we're seeing here too is that technology makes it possible to do these things. I think you can argue that as we think about the campaigns that we've experienced, particularly in Iraq (maybe more than Afghanistan), it is very easy to get fixated on a particular individual opponent and then to assume that once you got them then everything stops on the other side. That's not true. On the other hand, it is absolutely the case that we're in a world where, as in the past, the nature of the opposing commander makes a big difference. You might say that part of the Western way of war was to try to get past genius and the idea that it really all depends on who the general is. This takes us back to Clausewitz who says that 'You study genius not because you're going to become a genius but because you will understand something about the nature of war.' Well, maybe so.

PR: Those comments lead me to think whether warfare is being regarded – certainly by our military and politicians – as a science and not as an art: that the imponderable variables of human creativity, ingenuity and dynamism are being disregarded in favour of the more certain variables that hard data can prove or disprove in an binary way. As a final point, I wondered what your view was on the part that technology played [in war and combat] since you raised it right at the start. It strikes me from your last comment that what you've said is what technology makes possible is not necessarily decisive, let alone war-ending and we're now allowing technology to shape our approach to war. Do you think that is helpful or not?

EC: I think it's very hard to say whether or not it's helpful. My first instinct is just to figure out how exactly is it shaping what we do and the way war is waged. For me that's the first order question before I can get at whether it's helpful. I'll give you just two things that I find very striking about what technology is doing right now. One is if you look at the Nagorno-Karabakh war where the Armenians were basically wiped out by the Azeris backed by Turkish drones: there you have a middle power using a technology [attack drones] which was pioneered in the West. It was extraordinarily effective against an industrial era military. We've seen somewhat similar things in Libya. What does that mean? I'm not quite sure but it certainly indicates that something is very, very different. We need to figure it out. The second thing (and there's only been a bit about this that has come out about the Gaza war), is that I think it's also actually pretty important is the extent to which the Israelis were using AI to generate targets and then to act on them integrating many, many different sources of information in very compressed time periods. That would seem to indicate that what we may have are forms of warfare where the cycles of identifying targets and striking them are now getting much more compressed but also a lot more automated with all the perils of that.

Again, I don't think one can say a priori whether that's a good thing or a bad thing, helpful or unhelpful from the point of view of the United States or the UK but I do think those are the phenomena that one really should be thinking about very hard in places like RUSI or, dare I say it, SAIS.

Reflections

John Stuart Mill used the phrase, "The slumber of decided opinion" for a different context. However, there is something within that quote that instinctively applies to national security discussions, particularly when the military try to have the dominant voice. Given the peculiarities of the deeply imbedded hierarchical structures of today's Western military organisations, the dangers of military power-distance relationships and epistemic communities, the dangers of allowing militaries to revert to an era of lazy thinking based solely on their experience, and the denigration of any view that is counter to their own biases and assumptions, are significant.

Some of those assumptions and presumptions have been brought about by the lack of exposure to conflict in Western states. A very small proportion of Western societies have been exposed to the violence and destruction from military interventions since 1989. It has also been a very small proportion of the military force that gained that understanding. The 'decided opinion' of contemporary military thinking is built on constabulary operations, counter insurgency and counter terrorism campaigns that bear little resemblance to the high intensity fighting from which adversaries are drawing lessons. Whilst enemies are constantly thinking about the dynamism of combat, Western militaries have been seduced into a world governed by mythical end-states, measurable effects, and a continuing fabled belief in the inspiring general.

In reflecting on the conversation with Professor Cohen there are identifiable lessons that have not been addressed in the four years since the discussion was recorded and the writing of this piece. Still today, and without learning anything from recent history (and their own experiences of failed military campaigns) Western politicians continue to sign off inarticulate and poorly designed military plans for action without posing the fundamental questions. As Eliot (and others) noted, these questions can be simply phrased but complex and difficult to answer. They cannot be fitted into a social media post; they cannot be summarised into a one-page brief or a PowerPoint slide. Yet without addressing these questions, there seems little hope that the West will rediscover a way of thinking about war and warfare that enables military forces to be successful. That is something we really need to be deeply concerned about.

The Western Way of War
Can We Draw Conclusions?

In reflecting on the conversations transcribed in the previous chapters, and in the other – no less worthy – hours of conversation about the Western way of war, a number of key questions emerged. Rather than attempting to apply a reductionist frame to all the discussions and draw some key trends or patterns, it is to these questions that this conclusion applies itself. The answers to the questions are a very personal set of reflections: in this, the views, opinions and weights remain the author's rather than those of contributors.

Is There a Dominant West in Intellectual Terms?

In terms of military thinking, the dominant discourse, ideas and concepts come from the US military. Today, it seems that most European states simply adhere slavishly to the latest product from US Army Futures Command. In recent history there was a noticeable influence on US military thinking from the other Five Eyes nations (UK, Canada, Australia and New Zealand). After the Cold War, these states – specifically their militaries – were instrumental in shaping US doctrine, in Europe if nowhere else. However, the US military's intellectual renaissance post-Vietnam was far more influenced by analysis of the 1973 Yom Kippur War than any influence from European Allies. There is scant evidence of any European thinking in the emergence of AirLand Battle as a US concept in the 1980s; thereafter, Europe's place in the development over a Western way of war is hard to find.

However, the US military appears to be striving for adherence to development protocols instead of original thinking. US military alignment across services on topics like Multi Domain Operation (MDO) or Joint All Domain Command and Control (JADC2) has over-taken any attempts to develop a cadre that would think more deeply about *how* US forces might be required to fight. This becomes deeply concerning as adversaries start to adapt to US planned ways of fighting, and diverging from a single way of war that Western forces could face in the coming decade. If one examines the US military response to the failures of manoeuvre doctrine in Ukraine in 2023, one sees a narrative that states the issues as one of 'failure to adhere to processes and strict protocols' rather than an acceptance that a single methodology of fighting might just not work against any and all adversaries.

Those that do acknowledge the requirement for differentiation and unpredictability are also deviating from the prescribed US way of war. Both in India and Israel, military leaders, political masters, and military forces are actively pursuing different paths on *how* to fight. None of these might be successful, but the divergence from US core concepts is noteworthy.

Today, as US military thinking seems to be bogged down in technicalities rather than intellectual advancement, Australia – of all Western states – is demonstrating the most interesting, advanced, and rigorous conceptual challenge to the US from serving personnel. After decades of watching NATO from the side-lines, Australian military professionals now seem to stand head and shoulders above their European colleagues in terms of intellectual curiosity and ingenuity.

Peculiarly, Europe retains some of the finest minds in thinking about war and warfare, but these individual personalities are spread across European academic institutions and lack influence on either US thinking or, strangely, their own state militaries. So while there are smart people thinking in interesting and nuanced ways, they are individuals and their collective wisdom holds little sway.

Are the Prisms for Examining the Western Way of War Too Limited?

A critique often levelled at the discussions about the Western way of war is that active participants are drawn from a small community of interest. It is true that few Russian or Chinese speakers address the topic, but the podcast series did manage to elicit interviews with personnel from Israel, Georgia, India, Norway and Chile. The narratives from the Middle East or the Levant country hold a different perspective from those in Europe or other Western states. As these countries have been at the centre of regional battlegrounds for the West in the past 20 years they, clearly, have views on the Western way of warfare that suggest that the Western way of warfare should be characterised as cowardice: constantly developing technology to avoid boots on the ground and shying away from a human imperil to Westerners: an attempt to achieve uncertain political ends without risk or investment.

In seeking to balance some of these views, evidence was sought from Western based experts of these states to reflect such views. This approach was a useful period of difficult self-reflection on Western performance as viewed by others, whether Persian, African, Asian or Latin American. Hearing such accounts is often painful to those so deeply embedded in Western militaries, yet it is necessary. The aim should always be to promote critically self-reflective thinking rather than celebratory (an approach most of those who have experienced combat operations adhere to) and is an important part of understanding perspectives on how to fight in order to adapt and improve.

The objective of our assessment of the performance of Western military forces, through *how* they fight (and have fought), and *how* they aspire to fight, was in attempting to understand how this way of war might react when faced with adversaries fighting in very different ways. One of the key points to emerge is that adversaries and potential enemies envisage fighting in very different ways from those we expect and depend on: their behaviours and rules for the battlefield reflect cultures, ethics and morals very different to the homogenised Western variants that have been lauded by successive leaders in Western capitals.

Moreover, adversaries have demonstrated an ability to interrogate Western military doctrine and political ambition and adapt around it, both technologically and tactically. The ideas, for example, that either decapitation strategies and battlefield manoeuvre are preordained methods to achieve victory on a battlefield have been debunked on battlefields from Ukraine to Mali, from Libya to Yemen. Even after witnessing first hand their failures in Iraq and Afghanistan, Western political and military leaders continue to double down on these concepts. It is concerning that the lack of evidence to support their utility is never considered prior or between campaigns, but more worrying, perhaps, is the lack of alternative strategies being discussed by those leaders – even if it was only to examine differing options and then deliberately disregard them as unsuitable for the fight or the context.

Does the Western Way of Warfare Actually Exist?

The core question that should be answered in looking at this topic is to wonder whether a Western way of war actually exists at all. Is there a supreme art (or science) to wielding the military instruments by Western, liberal, capitalist democracies? Does the historical record indicate an enduring set of characteristics in the way Western states fight that still survives? Is it true, as per Victor Davis Hanson, that our fictional belief in dismounted close combat is based on a deep-seated desire to be immortalised as those from ancient Greek city states were? Is there something within the Western profession of arms that yearns to be a modern Archilles or a Christian knight from the crusades? And what of the discarded, less heroic strategies of Hector or Odysseus?

These questions are valid; indeed, they are important in situating the entire discussion about a Western way of war, and in providing utility in the discussions for future members of the profession of arms. It should also be accepted that combat is the same experience for a lot of people, and not just the civilians and the kids in conflict zones. This is about the soldiers, sailors, air force personnel, marines who are doing the fighting. All of them experience close combat in much the same way. The fear, the dread, the adrenaline. Afterwards, there's the post-traumatic rush, the stress that comes out in how we feel about it. Much of the emotion, the physicality, and the brutal nature of fighting is the same; as is the training for combat operations across militaries (whether Western or otherwise): Drill is much the same in most countries. Some armies might goose step rather than march with high knees but in very broad terms the similarities are remarkable. Remarkable but perhaps not surprising. The really broad themes about militaries, how they train, how they are organised, what weapons are favoured, are all functions of an evolution in humanity's desire to compete and wage war. Militaries are constructed for a purpose, not by accident, and they are employed to gain something from fighting.

Despite those similarities, there has been change and adaptation. In *Perilous Glory*, John France challenges readers to take a long view of warfare and try to understand how it has been shaped.[1] In reflecting on France versus the narrative of Hanson, it is less than compelling to subscribe to a belief that we are still all recognisable as Spartans. Western militaries don't

---

[1] John France, *Perilous Glory* (Yale University Press, 2011).

fight in phalanx formation, we don't have the same belief systems, ethics, morals, societal constructs as they did. We have evolved, as has the way we wage war.

But there is utility in examining ancient history, specifically in thinking about the diametrically opposed ways of fighting as personified by Hector, Archilles, and Odysseus. This will be a sweeping generalisation but has some truth in it: The West wants to fight like Hector, full of honour and ethics and love and the people on the defensive all the time, mastering their skills but protecting their resources. A farmer by day, a governor of the people and yet metering out judgment and honour and values and ethics wherever they can. Indeed, the popular accounting for Western military actions portrays them distinctly in this light. It has become so emmeshed in the narratives about contemporary war that it has become heresy to consider waging war in another way.

Yet there are others that actually reflect the reality of combat on the ground. Some of the special forces people who have become Achilles, the uber warriors, the fighters, the slayers and bringers of death and destruction, they are an essential part of the Western way of war as any portrayal of Hector. You go into battle and frankly, you don't want to kill everyone but actually, sometimes, that is the outcome that's going to happen and you train for that and really make it happen. The people who can deliver those things on a battlefield are vital to how you wage war.

Then you have Odysseus who personifies strategies that actually are the only way that many smaller Western countries have found themselves able to leverage their own capabilities, their own strategic culture, and the way they feel about combat. In a previous age it was through cunning, guile, wit and deception that was a core part of how Britain won wars. That was not necessarily a deliberate strategy selection to start with, and it is certainly not how modern history imagines it, but an accounting for British military success does not reveal a series of pitched fair-fights with honourable objectives. Indeed, by ending up with this way of fighting – outside the established rules of warfare – Britain was most successful.

There is, of course, a real tension between what the politicians want (which is honourable, rules-based, respectful strategies of Hector), and what the military want in terms of ending a war as fast as possible, even if that means breaking rules and using extremes of violence (which is probably the Achilles version). When reality bites, however, forces often find themselves being in a fight and having neither of the requirements or permissions to make either of those strategies a success, nor with enough Achilles-like capabilities to win. The result occasionally ends up with an Odyssean approach in order to bring about success in the end.

None of that is to circumvent the core question: Is there a single Western way of war? There are, perhaps, some broad conclusions that should be expressed. First, it is clearly evidenced that Western militaries (and their political leaders) like to hang their hat on something: a central concept around which it can galvanise the national security community. This might be information or digital backbones, data or technology, silver bullets or special forces, ethics or morals. In broad terms this is not so dissimilar to how others behave: conversations about the military instrument cross over in this regard with China, Iran, Russia, or North Korea. All states are examining the utility of warfare as a means of achieving their national interests in protecting and ensuring their national identity, their purpose, their culture. Indeed, all these states will talk about their respective values too, although this might be better expressed as national interests.

Given these significant commonalities across states one might disregard the idea of a Western way of war. However, there are – certainly in terms of contemporary military operations – some key aspects about *how* the West wages war and intervenes militarily that does feel distinctly Western. Some have classified it as a uniquely expeditionary outlook, or related it to temporal limitations (go in, go hard, go fast, go home). There are classifications about the technological reliance of a Western way of war, an idea that it is about finding the cheapest possible force, or even an idea that it is a necessity to lose the first battle and then winning in the long run. Of all these characteristics, the one thing you can definitely identify as Western more than any other fighting style that we've looked at, is about the West's ability to adapt during the fight it is in. Many great commanders and analysts have said that adaptability in the face of a battle is the most critical element that can drive success on the battlefield, primarily because, as an organisation, you are never prepared to walk into the fight that you're going to have.

As all of those preconditions, caveats and exceptions became evident, it was apparent that a modern Western way of war did exist; one led by US doctrine and an American concept of fighting. This conclusion has been less about a tipping point or event, but rather the accumulation of evidence from relevant literature, from august speakers, from observations of military actions and campaigns, and from other conversations, whether it's in France, in Moscow or with people from Nanjing or Tehran. There is a distinct feeling, to Westerners and around the world, that that there is a Western approach to war that is recognisable, and is recognised, understood and acknowledged.

There is a sense that the coalescence of the Western way of war happened in NATO in the shadow of World War Two. The merging of American, French, German, British and Imperial fighting doctrine was not an overnight agreement or alignment. Indeed, it was not until the arrival of AirLand Battle doctrine from the US in the 1980s that the threading of military philosophy across Western states really emerged. Itself the result of an intellectual renaissance of US military thinking after the defeat in Vietnam, and stalemate in Korea, as well as a clear-headed assessment of the 1973 Yom Kippur war, US military personnel formed a theory of war and deep battle founded on the needs of the context rather than a reliance on their own cultural experiences. It was a concept forged from the welding of Russian deep battle, combined arms manoeuvre, precision, airpower, nuclear weapons, and production rations alongside an appreciation of time, space, the adversary and context. The underlying issue, however, was the idea that commanders and troops would not be able to communicate effectively during a war, given the fragility of communications networks and the contested electronic warfare environment: the commander on the ground would have the best understanding of the battle requiring a philosophy of delegated command and responsibility to be embraced across the Alliance military structures.

Thereafter, perhaps because it was perceived to be the reason for the downfall (defeat in military terms) of the USSR, the ideas of AirLand Battle doctrine became the single Western approach to war. Evidenced in Kuwait, Bosnia, Kosovo, Sierra Leone, Iraq, and Libya, this approach of war – a Western way of war – went unchallenged as *the* way forces would fight. It became so ubiquitous in campaigns no matter the result (successful or otherwise) that few political leaders or Western military officers can even imagine an alternative.

However, there was a more worrying trend that also emerged. Whether in Georgia, Nagorno-Karabakh, Libya, Yemen, or Syria, a whole host of contemporary conflicts started to expose behaviours and tactics by belligerents that made for interesting analysis. It was becoming increasingly clear that Western forces would not perform well against sub-peer adversaries in these conflicts, despite the disparity in capabilities, training, connectivity, intelligence, and technology. Few would listen to an argument that evidenced the performance gap between Western forces and modern combat conditions. Despite Western military change programmes, modernisation, and transformations, there was little evidence that the West could survive contact with the way adversaries were fighting. It was the dependence on some of the big-ticket items that the Western militaries had been very publicly investing in (data, digitisation, technology, cyber, space), and against which adversaries developed ways and capabilities to overcome and bypass.

These belligerents had been carefully studying Western (US) military operational techniques, operational and organisational design, and tactics since the 2003 Shock and Awe campaign in Iraq. By the time of the second Nagorno-Karabakh campaign (2020), adversaries had identified the Western way of war as being centralised and highly reliant on signals, data, information and orders. As a result, adaptation to such conditions reversed the dilemma and began using decapitation techniques against Western trained headquarters. Moving a step further, belligerents had such a good understanding of the electronic elements of the battlespace (a core requirement in fighting a highly centralised adversary from the West) that they could start employing a recce-strike complex against individuals.

As a result, we saw that as soon as you emit on a radio, or a radar, at any level from a headquarters down to an individual unit (a single tank, an armoured vehicle, an infantry patrol), that call for fires and artillery shells and drone launch rockets, they came onto your location rather swiftly. The new rule of contemporary warfare was that to transmit on the battlefield spells death. This kind of conclusion did not reflect well on the Western idea of a connected force: the extension of the AirLand Battle concept. Similar conversations were had about drones (Libya 2020), armour (India, 2019), lethality (Mali, 2017), and survivability (Sweden, 2018), about urban warfare (Iraq, 2016), the limitations of manoeuvre as a single doctrine (Yemen, 2018), and the dependency on precision navigation (Baltics, 2017). Few of these facets were accepted in Western capitals by either military or political leaders.

Given the reticence (bordering on hostility) from Western military personnel to critiques about their way of war evidenced by discussions between 2014 and 2021, the important corollary to the core questions (does the Western way of war exist?) is whether that concept of a modern methodology about *how* Western militaries fight has adapted, deliberately or otherwise since the 1980s?

The ambition to change and adapt has been in the lexicon of military forces since 1990. Although wielded in speeches more for the value of narrative and an aspiration for relevance, significant changes – rather than tweaks – to the core concept of the Western way of war has little supporting evidence. The endless claims of revolutions in military affairs (of one form or another) have, in fact, had little impact on how Western militaries have been organising and preparing to fight. The rhetoric from successive military leaders seems to herald another transformation on an almost weekly basis, and usually related to the latest technology promised by a commercial enterprise. It is exceptionally rare for such promises to deliver anything, let alone something that has an impact on fighting. Generals became more ebullient about technology in the 2020s, a domain of interest previously reserved for senior naval and air force personnel. The 'connected' force became a mantra for Western armies, following a US ambition to deliver greater lethality to every soldier.

Despite the hype, however, there seemed little change in real, valuable change in evidence. Tactics, organisation, training and education remained very fixed to a post-Cold War era, distracted by the promises of the prevalence of surrogate warfare (Krieg, 20190, Fourth Generation warfare (Rid, 2009), Proxy warfare (Hughes, 2012), Hybrid warfare (Hoffman, 2007), Non-linear warfare (Galeotti, 2016), Grey Zone warfare (Echavaria, 2016), or Shadow warfare (McFate, 2019). These conceptual distractions coupled with the emergence of technology as a perceived driving force for change was sufficient for most military leaders to forget about people, creativity and fighting at close quarters.

Over this period, it was usual to hear military commanders acknowledge the importance of people in anything other than an aside. It came as a welcome surprise, therefore to hear General McConville, the US Army Chief of Staff, talk about these big recapitalisation programmes and digitisation and at the end, hear him say, "but all that doesn't matter because it's about people at the end of it".

As the West entered 2021, some leaders started to accept a new era was perhaps arriving, driven less by their obsession with promises of technological overmatch, and more by the behaviour and capabilities of potential enemies. USMC Commandant General Dave Berger developed an entirely new fighting doctrine for the USMC based on the geography, topography and geometry of the potential Pacific campaign against the Chinese forces of the People's Liberation Army.

Other deliberate changes to the Western way of war being postulated today are concerning however, primarily because they seem to intimate that a single modality of warfare can work against any and all adversaries. The central facet to that new doctrine of multi domain operations is the connected force. That singular idea – fast becoming a dependency on which disproportionate effort is being placed – is particularly worrying at the moment as narratives dictate that these changes are caused by Western militaries adapting to something new in adversary's behaviour. In fact, it is more of a linear pathway back to AirLand Battle, but without the understanding of context, geography and adversary. It also brings with it a centralisation, a reliance and a dependency on data and communications for a connected force. This should be raising red flags with military personnel because we know that adversaries have been building militaries and operating in campaigns designed to bypass and exploit such a reliance.

That centralisation of military command and control is also diametrically opposed to the original concepts of mission command, delegation, and empowerment. That might be the right shift for the fights that Western militaries will engage in over the course of the coming decades, but is not being recognised, acknowledged or articulated. From everything we know about human behaviours on the battlefield – whether Western or differing ways of war – there does not appear to have been cases where both mission command and centralisation can co-exist in a productive and effective way: the drivers – indeed, the philosophies – for each are diametrically opposed. Technologists have promised that the centralisation of command and control – if they are to exist as entwined entities – will provide a new, unprecedented level of understanding control and efficiency to Western militaries. If they are correct, centralisation might become a battle winning approach for Western forces. But this is neither how they are training now, nor what their doctrine is espousing for the future.

Warfare and Technology

Many authors, practitioners and scholars are promising nearly instant victory through sophisticated technology. Whether in ideas like 'Hyperwar',[2] to the trend of techno-fiction defining military requirements and concepts,

---

[2] Amir Husain et al., *Hyperwar: conflict and competition in the AI century* (AM Press, 2018).

the promise that technology will enable decisive victory at a much-reduced human and financial cost to societies. These are strange claims to be making based on contemporary experiences of war and warfare. In Iraq and Afghanistan, the US led Western coalitions had generational advantages in terms of technology and mass. The same was true for the French forces in Mali, and – to a large extent – the Ukrainian forces fighting the Russian army in Ukraine. Technology – whether in one area or many – does not bestow on the employer a preordained right to victory.

There is an additional danger that runs alongside the current narrative of technology and warfare: that the promise of cheap and easy victories starts to separate war from its political aims by making military victory an end in itself. Yet this kind of discourse fails to acknowledge that war is human and political, therefore, highly uncertain. The consequences of starting to imagine war without those facets would be terrible.

Technologies (new or – more usually – rediscovered ones like electronic warfare) are important in military operations. But these are tools and capabilities that do not represent concepts of warfare or a strategy, plan or philosophy with which to engage in warfare.

Few people who work in a modern military – particularly those involved at the sharp end of combat operations – have an experience with technology that rarely lives up to the promises of commercial providers, military or political leaders. New equipment is rarely delivered on time, on budget, or with the promised performance. New tanks rarely have enduring unmatched superiority against an adversary; belligerents have a nasty habit of inventing ways of overcoming any temporal advantage of an opponent. New surface-to-air missiles might work well but are too expensive to be used against multiple targets. The software system fails to work on actual artillery pieces (as opposed to a simulator), or maybe the latest software patch for an aircraft causes an entire fleet to be grounded. Because war is a reciprocal dynamic, adversaries are inventing, adapting, innovating and changing form at the same time. So, the idea that somehow one single technology like Artificial Intelligence will change warfare fundamentally, is well meaning (and often a compelling sales pitch) but rarely delivers in the way intended. The same conversations and promises were evident during the golden age of cyber warfare discussions.

Have We Been Honest and Open about the Western Way of War?

There is a valid critique that, at the political level, the effectiveness of the Western way of war has been over exaggerated. It has been argued that post World War Two interventions by Western coalitions led by the US have not been successful, but the political narrative has not been honest with the public perhaps even dishonest with the force elements that have been used. How far should that critique go? Is it possible that the use of force has become extremely ineffective given our current approach to warfare?

The idea that Western interventions haven't been successful is evidentially clear; most of the intelligent debate admits failures in Iraq, Afghanistan, Libya and more. Yet there are some things that Western militaries are doing well at. The tactical level fighting by individual units has been exceptional and has grown out from the combat experience many states have had since 2001. However, the military approach has failed at the operational level of war, at the brigade, divisional level of command and at the corps level. Western military success has not been found at the multi-national force level: indeed, it is that interaction between states that military integration has had less than optimal progress. Western military interventions have also failed politically and economically; the promises to bring Russia to its knees with sanctions after their invasion of Ukraine (again) has not delivered success. All these things are true.

Yet, despite this acknowledgement by people who have left office that we have not succeeded, it is troubling that we have not amended a great deal of our military organisations, structures, processes, or education as a result. If there is no hubris exhibited about the Western performance in recent campaigns, neither has there been a drive towards self-reflection and change. The questions that need addressing should be coming across the levels of military and political operations: How do we change operational planning so it doesn't just run on a six-month cycle of rotations? How do we make the core level structure appropriate? Do we need to think about the balance of centralisation again? How do we move away from the dependencies that we have in warfare that make us fragile and undermine resilience? How do we train political leaders to make better decisions? How do we move away from giving politically aware military advice to just giving military advice and let the politicians decide what they want to do? How do we move away from the heresy's and ill-found assumptions we have about casualties or risk appetites or blood money or interests? How do we move politicians away from the ridiculous conversations that they have and statements they make about existential fights and critical national interest and values where actually, they don't really mean those things at all?

The lack of useful conversations that drive a better dialogue represents, in effect, an ambivalence towards both previous Western interventions and the current military instrument. And that ambivalence is by politicians and military personnel too.

It is perhaps the case that the root cause of this malaise is the absence of mature and honest conversations about national security both within Western capitals, and with societies more widely. It is rare to hear a discussion that seeks to understand both what politicians would want to achieve with the military instrument, or what the military – as currently funded – can actually do. Without these during times of 'peace', our ability to put these into practice during periods of tension, friction, threat, or war becomes increasingly unrepresentative of reality.

Have National Ways of War Disappeared?

The basis for this work is an assumption that various Western schools of war had been homogenized into a single Western way of war. Many scholars have pointed at NATO as a shredding machine for sovereign national doctrine. Some critics who use France as evidence for that; point out that when France made the decision in the 2010s to align its doctrine to NATO STANAGs and directives, the French art of war has but died. There are also critiques that French PME, for example, has become merely a drilling machine for comprehensive operational planning directive exercises.

As an example, the various principles of war – once highly differentiated between states and their schools of war – have become aligned towards a single Western interpretation. Historically, indeed up to the end of the Second World War, the French, British and Prussian principles were very different. The alignment took time but did occur through the auspices of NATO. It would be challenging to attribute this homogenization as a deliberate organisational objective but rather a military acknowledgement of the principles needed to fight a known adversary (the forces of the USSR in Western Europe).

This distinction is important. Those who have worked with various US forces would be able to identify the differences between those that were based on the US East coast (aware of and largely abiding by NATO operating principles), and those based on the US West coast (leaning more towards interoperability with forces and partners in Korea, Japan, and against a different series of military threats and operating patterns).

It is possible, it seems, to maintain two schools of war even within a single national military. In an academic sense, this seems to be eminently sensible. A way of war – *how* a force plans to fight – must acknowledge the battlefield geometry and the adversary. Thus a way of war for the Pacific (facing China or Korea), will necessarily be very different from one facing Russia in the Arctic or Western Europe. Indeed, both of those methodologies for war might not be appropriate for fighting in a Middle Eastern theatre.

But while the US seems to have retained an ability to potentially differentiate, there is little evidence of this flexibility or intellectual dynamism in European schools of war. Perhaps, given the priority of their threats (Russia), this should not be an overwhelming concern; nonetheless, it is worth retaining as a presumption if NATO was to undertake another out-of-area operation or campaign.

Interestingly, while the West has been busy working towards a single operating concept, other states (and actors) have been diverging in the way they are planning to fight. Many states have identifiable ways of war evidenced both through fighting styles and their methodology of training and equipping their forces. So, while the West has been aligning, potential foes have been diverging.

How Central Are People to the Western Way of War?

As noted by many scholars and practitioners, indeed as far back as Thucydides, whatever way you look at it, when you try to define war people remain at the centre of the equation. There are truisms on war, which were as valid 2,500 years ago as they are today: people fight for three things – fear, honour and interest.

It would be foolhardy to ignore evidence, trends, patterns, behaviours, and reality to believe that war is not a distinctly human endeavour. The requirement for human imperil is one of the critical factors of the rubric that is warfare. Indeed, that requirement for human imperil and risk is one of the elements to make decision-making in war as considered as it is. Without that, perhaps the idea that war could be fought solely by machines for example, places a different set of perspectives in front of politicians and decision makers. If the only risk is financial (the cost of the machines), doesn't that make war more likely?

On fighting wars, John Boyd used the line, "People, ideas, equipment – in that order." One might argue about the relative priority order of people and ideas, but to anyone who has studied this in detail, both sit on a higher plane than equipment. Yet at the moment the West prioritises equipment, then people and doesn't think about ideas (and in fact, it doesn't really think about people that much either). This is an area that should cause us considerable concern.

Anyone exposed to the personnel of the current military forces in the West cannot doubt that they will adapt to whatever challenge put in front of them; they are trained reasonably well and they will try their damnedest no matter what kit you give them. Yet the concepts of how they plan to fight are based on some historical norms and selective referencing that is well published. That idea of how the West will fight is critical because others – whether Russia, China, Iran, North Korea, or even lesser actors like the Houthi's in Yemen or political factions in sub-Saharan Africa – all have very distinct ideas of how they are going to fight in the future.

The Western concepts of how to fight look a lot like what it did in Iraq (in 2003), or in Sierra Leone (in 2001) or in Bosnia (during the 1990s). Today's Western approach to warfare is a linear extension of those experiences. The amendments to that idea of *how* they will fight is an assumption that the Western forces will be beautifully connected and acting seamlessly as the promise of technology removes any possible 'fog of war'. There is no differentiation between how they will need address different adversaries – the differing geography, topography, behaviours, or approaches of differing adversaries.

It is a peculiar facet of discussions with Western political and military leaders that few doubt the centrality of people in warfare. War certainly has elements of the mechanical, of code, of information, of data, of logistics, medical, and engineering. However, all evidence continues to point to some distinctly human characteristics that deliver success: the ideas of rationality and logic, cognition and art, and creativity. If these provide the 'edge' that Western leaders are seeking as the panacea for 'Victory', it seems somewhat paradoxical that so little effort is focused on their development within the Western militaries.

The Pipe Dream of Easy War

Concepts of defence and warfare must consider social, economic and historical factors that constitute the human dimension of war. As HR McMaster put it, "But in the years preceding our last two wars, thinking about defense undervalued the human as well as the political aspects of war. Although combat operations unseated the Taliban and the Saddam Hussein regime, a poor understanding of the recent histories of the Afghan and Iraqi peoples undermined efforts to consolidate early battlefield gains into lasting security."[3]

There has been a predilection in Western societies (including their militaries), in the belief that the world order of 1991-2022 was an equilibrium that could be sustained by small numbers of technologically sophisticated forces capable of launching precision strikes against enemy targets from safe distances. Further, that this kind of remote warfare would be a sufficient deterrent to others in order to maintain this Western concept of 'peace' [sic]. The hubris of the arguments should be deeply concerning, as is the rhetoric being employed by successive military leaders in the West about 'Network Enabled Warfare', 'Effects Based Operations', 'Rapid, Decisive Operations', 'Shock and Awe', 'Full-Spectrum Dominance', 'Multi Domain Operations' [Integration], or 'Cross Government coherence'. All of these concepts were designed to perpetuate Western supremacy in military terms.

In accepting these guiding principles, militaries across the West – like their political masters – omitted to address the fundamental, first-order questions. As both Eliot Cohen and Mike Clark have noted, these questions

---

[3]   HR McMaster, 'The Pipe Dream of Easy War', OpEd, *New York Times*, 20 July 2013. https://www.nytimes.com/2013/07/21/opinion/sunday/the-pipe-dream-of-easy-war.html accessed 18 March 2024.

might not guarantee success, but they considerably reduce the probability of failure: an undeniable result of the Western led interventions since 1991.

Answering such questions is not easy, but the failure to address them by prioritising the intellectual attributes of the military machine to those around business process and financial accounting is not a way to buy out the core conceptual problems. If the grand concepts are not being challenged, there is little evidence that the core military questions are being addressed either. There appears to be a reluctance in Western militaries to admit that warfare is difficult: from a personal level when engaged in combat actions; the ideas of integrating multiple fires on a single or series of targets; right through to understanding why any campaign plan would work. Addressing the key assumptions in Western plans through challenging interrogations along the lines of 'What is our theory of change?' are not only worthwhile, they are essential. The current dynamic in Western capitals is to avoid these questions because they are hard: as a result it seems that not only are they not being addressed, but are being written out of our professional military education syllabus. Given these trends, it is entirely possible that future military and political leaders run the risk of never knowing they are important.

Bibliography

Aboudouh, A. "Where do the Taliban get their money and weapons from?", *The Independent*, 1 September 2021. https://www.independent.co.uk/asia/south-asia/taliban-where-weapons-money-funds-b1911655.html.

Alman, D. and H. Venable, 'Bending the Principle of Mass: Why That Approach No Longer Works for AirPower', *War on the Rocks*, 15 September 2020, https://warontherocks.com/2020/09/bending-the-principle-of-mass-why-that-approach-no-longer-works-for-airpower/.

Barno, D. and N. Bensahel, 'Are you enough? Our speech to the PME class of 2019', *War on the Rocks*. 18 September 2018, https://warontherocks.com/2018/09/are-you-enough-our-speech-to-the-pme-class-of-2019/.

'Blair Doctrine', https://archive.globalpolicy.org/component/content/article/154-general/26026.html.

Bratic, V. 'Examining peace-oriented media in areas of violent conflict', *International Communications Gazette,* vol. 60, no. 6 (2008).

British Defence Doctrine Ed6, https://assets.publishing.service.gov.uk/government/uploads/system/uploads/attachment_data/file/1118720/UK_Defence_Doctrine_Ed6.pdf.

British Defence Doctrine Ed5, https://assets.publishing.service.gov.uk/government/uploads/system/uploads/attachment_data/file/389755/20141208-JDP_0_01_Ed_5_UK_Defence_Doctrine.pdf .

Brown, GC. 'Understanding the Risks and Realities of China's Nuclear Forces', in Arms Control Association commentary, June 2021, https://www.armscontrol.org/act/2021-06/features/understanding-risks-realities-chinas-nuclear-forces.

'Costs of War', Watson Institute of International and Public Affairs, Brown University. https://watson.brown.edu/costsofwar/figures/2021/human-and-budgetary-costs-date-us-war-afghanistan-2001-2022.

Connolly, WE. *The Terms of Political Discourse* (Princeton University Press, 1983).

de Pizan, C. and A. Kennedy, *The Book of the Body Politic* (Iter Press, 2021).

Economist, "How Ukraine is using AI to fight Russia", 8 April 2024. https://www.economist.com/science-and-technology/2024/04/08/how-ukraine-is-using-ai-to-fight-russia.

Economist, "How Missiles are changing the Middle East", 22 November 2023. https://www.economist.com/films/2023/11/22/how-missiles-are-changing-the-middle-east.

Fox, A. 'Reflections on Russia's 2022 Invasion of Ukraine', Land Warfare Paper 147, Association of the United States Army (AUSA), September 2022.

France, J. *Perilous Glory* (Yale University Press, 2011).

Freedman, L. "Alliance and the British Way in Warfare." *Review of International Studies*, vol. 21, no. 2, 1995, 145–58. JSTOR, http://www.jstor.org/stable/20097403.

French, D. *The British Way in Warfare, 1688-2000* (Unwin Hyman Inc., 1990).

Galeotti, M. 'I'm Sorry for Creating the 'Gerasimov Doctrine'', *Foreign Policy Magazine*, 5 March 2018. https://foreignpolicy.com/2018/03/05/im-sorry-for-creating-the-gerasimov-doctrine/.

Gallie, WB. 'Essentially Contested Concepts,' *Proceedings of the Aristotelian Society*, vol. 56 (1955/56).

Gat, A. *A History of Military Thought: From the Enlightenment to the Cold War* (Oxford University Press, 2011).

Gladwell, M. *Blink: The Power of Thinking Without Thinking* (Black Bay Books, Little Brown, 2005).

Global Strategic Trends (2018), Defence Concepts and Doctrine Centre, UK Ministry of Defence, 2018, https://assets.publishing.service.gov.uk/government/uploads/system/uploads/attachment_data/file/1075981/GST_the_future_starts_today.pdf.

Haines, JR. 'How, Why and Where Russia will deploy Little Green Men – And Why the US Cannot', Foreign Policy Research Institute, 9 March 2016, https://www.fpri.org/article/2016/03/how-why-and-when-russia-will-deploy-little-green-men-and-why-the-us-cannot/.

Hammond, M. *Marcus Aurelius: Meditations* (Penguin Classics, 2006).

Hanson, VD. *The Western Way of War: Infantry Fighting in Ancient Greece* (University of California Press, 2000).

Havrén, SA. 'China's No First Use Policy: Change or False Alarm', in RUSI commentary 13 October 2023, https://www.rusi.org/explore-our-research/publications/commentary/chinas-no-first-use-nuclear-weapons-policy-change-or-false-alarm.

Hawkins, M. *The Power of Boredom* (Cold Noodle Creative, 2016).

Hoffman, F. 'Conflict in the 21st Century: The Rise of Hybrid Wars', Potomac Institute for Policy Studies (Arlington, Virginia), December 2007, https://www.potomacinstitute.org/images/stories/publications/potomac_hybridwar_0108.pdf.

Hoffman, F. "A Second Look at the Powell Doctrine", *War on the Rocks*, 20 February 2014, https://warontherocks.com/2014/02/a-second-look-at-the-powell-doctrine/.

Holden Reid, B. 'Introduction: Is There a British Military 'Philosophy'?,' in JJG Mackenzie & Brian Holden Reid (eds.), *Central Region Versus Out of Area: Future Commitments* (Tri-Service, 1990).

Homer, BK. (ed.), *The Iliad* (Penguin Classics, 1992).

Howard, M. *Continental Commitments: Dilemma of British Defense Policy in the Era of the Two World Wars* (Prometheus Books, 1972).

Howard, M. 'The British Way in Warfare: A Reappraisal', in Michael Howard (ed.) *The Causes of Wars and Other Essays* (Harvard University Press, 1983).

Husain, A., et al., *Hyperwar: conflict and competition in the AI century* (AM Press, 2018).

Jash, A. "By the numbers: China's nuclear inventory continues to grow", Lowy Institute, 27 February 2024. https://www.lowyinstitute.org/the-interpreter/numbers-china-s-nuclear-inventory-continues-grow.

Kahnerman, D. *Thinking, Fast and Slow* (Penguin, 2016).

Kellner, D. 'Bushspeak and the Politics of Lying: Presidential Rhetoric in the 'War on Terror." *Presidential Studies Quarterly*, vol. 37, no. 4, 2007, 622–45. http://www.jstor.org/stable/27552281.

Lambert, A. 'The Naval War Course, Some Principles of Maritime Strategy and the Origins of 'The British Way in Warfare', in Keith Neilson & Greg Kennedy (eds.), *The British Way in Warfare: Power and the International System, 1856- 1956: Essays in honour of David French* (Ashgate, 2010), 251.

Liddel-Hart, B. "Economic Pressure or Continental Victories", *RUSI Journal*, vol. 76, no. 503 (1931).

Macmillan, A. 'Strategic Culture and British Grand Strategy 1945-1952', Aberystwyth University.

'Maritime Maneuvering and Tactical procedures", released in 1996, http://nato.radioscanner.ru/files/article66/1000.pdf.

Martin, M. *How to Fight a War* (C Hurst and Co, 2023).

McInnes, C. *Hot War, Cold War: The British Army's Way in Warfare 1945-1990* (Brassey's, 1996).

McMaster, HR. 'The Pipe Dream of Easy War', OpEd, *New York Times,* 20 July 2013. https://www.nytimes.com/2013/07/21/opinion/sunday/the-pipe-dream-of-easy-war.html accessed 18 March 2024.

McNeil, W. "What We Mean by the West", Western Civilisation in World Politics, *Orbis,* Fall 1997, 513-524, https://www.fpri.org/wp-content/uploads/2016/07/WH-McNeil-What-We-Mean-by-the-West.pdf.

Mercier, A. 'War and the Media: Consistency and Convulsion', *International Review of the Red Cross,* vol. 87, no. 860 (December 2005), 629-659. https://www.corteidh.or.cr/tablas/a21918.pdf.

Mill, JS. *On Liberty* (John Parker and Son, 1859).

NATO EXTAC 1000 series 'Maritime Maneuvering and Tactical procedures", released in 1996 http://nato.radioscanner.ru/files/article66/1000.pdf.

NATO ATP-3.2.1 "Allied Land Tactics", https://moodle.unob.cz/pluginfile.php/98398/mod_resource/content/1/ATP-3.2.1%5B1%5D%20ALLIED%20LAND%20TACTICS%202009.pdf.

Pinker, S. *Enlightenment Now: The Case for Reason, Science, Humanism, and Progress* (Penguin Books, 2018).

Rauta, V. and S. Monaghan, 'Global Britain in the grey zone: Between stagecraft and statecraft', *Contemporary Security Policy*, vol. 42, no. 4 (2021).

Raymond, GV. *Thai Military Power: A Culture of Strategic Accommodation* (NiAS Press, 2018).

Sankar, S. "Data is the new snake oil", *St Gallen Business Review*, 9 May 2019, https://www.stgallenbusinessreview.com/data-is-the-new-snake-oil/.

Skinner, D. "Air Land Battle Doctrine", Center for Naval Analyses, Professional Paper 463, September 1988, https://apps.dtic.mil/sti/pdfs/ADA202888.pdf.

Snyder, JL. *Soviet Strategic Culture: Implications for Limited Nuclear Operations* (RAND Corporation, 1977).

Storr, J. *Hall of Mirrors* (Helion and Company, 2019).

Strachan, H. 'The British Way in Warfare Revisited,' *The Historical Journal*, vol. 26, no. 2 (1983).

Strange, J. and R. Iron, 'Center of Gravity: What Clausewitz Really Meant', in *Joint Force Quarterly*, no. 35, 20-27. https://apps.dtic.mil/sti/pdfs/ADA520980.pdf.

Taylor, M. and M. Kent, "Media in transition in Bosnia: From propagandistic past to uncertain future", *International Communications Gazette*, vol. 62, no. 5 (2000).

US Department of Defense, *DoD Dictionary of Military and Associated Terms*, US DoD, November 2021, 30. https://irp.fas.org/doddir/dod/dictionary.pdf.

Venable, H. 'Decisive Maneuver is the Army Equivalent of the Air Force's Historical Emphasis on Strategic Attack as "The" Answer', *Linkedin*, 3 November 2023, https://www.linkedin.com/pulse/decisive-maneuver-army-equivalent-air-forces-emphasis-heather-venable-fzkhf.

von Clausewitz, C. *On War*.

Walker, P. and P. Roberts, *Wars Changed Landscape?* (Howgate, 2023).

Waltz, K. *Foreign Policy and Democratic Politics: The American and British Experience* (Longman, 1968).

Watkins, A. 'One Year Later: Taliban Reprise Repressive Rule, but Struggle to Build a State', United States Institute for Peace, 17 August 2022, https://www.usip.org/publications/2022/08/one-year-later-taliban-reprise-repressive-rule-struggle-build-state.

Widen, JJ. *Contemporary Military Theory* (Routledge, 2013).

Wintour, P. "UK could contribute to nuclear shield if Trump wins, suggests German minister", *The Guardian*, 15 February 2024. https://www.theguardian.com/world/2024/feb/15/uk-europe-nuclear-shield-donald-trump-germany-nato-deterrent.